SYSTEM-ON-A-CHIP VERIFICATION

Methodology and Techniques

SYSTEM-ON-A-CHIP VERIFICATION

Methodology and Techniques

Prakash Rashinkar
Peter Paterson
Leena Singh
Cadence Design Systems, Inc.

KLUWER ACADEMIC PUBLISHERS
Boston / Dordrecht / London

Distributors for North, Central and South America:
Kluwer Academic Publishers
101 Philip Drive
Assinippi Park
Norwell, Massachusetts 02061 USA
Telephone (781) 871-6600
Fax (781) 871-6528
E-Mail <kluwer@wkap.com>

Distributors for all other countries:
Kluwer Academic Publishers Group
Distribution Centre
Post Office Box 322
3300 AH Dordrecht, THE NETHERLANDS
Telephone 31 78 6392 392
Fax 31 78 6546 474
E-Mail <orderdept@wkap.nl>

Library of Congress Cataloging-in-Publication Data

A C.I.P. Catalogue record for this book is available
from the Library of Congress.

Printed on acid-free paper.

Printed in the United States of America

Contents

Authors

Prakash Rashinkar has over 15 years experience in system design and verification of embedded systems for communication satellites, launch vehicles and spacecraft ground systems, high-performance computing, switching, multimedia, and wireless applications. Prakash graduated with an MSEE from Regional Engineering College, Warangal, in India. He lead the team that was responsible for delivering the methodologies for SOC verification at Cadence Design Systems. Prakash is an active member of the VSIA Functional Verification DWG. He is currently Architect in the Vertical Markets and Design Environments Group at Cadence.

Peter Paterson has over 20 years experience in ASIC and computer systems design. Peter graduated with a BSEE from Robert Gordon's University in Scotland. He lead the teams that produced the first "mainframe-on-a-chip" (SCAMP) and single chip GaAs processor while at Unisys Corporation. These devices were early precursors to today's SOC devices. While at Cadence, he architected the platform-based SOC design methodology delivered to Scottish Enterprise as part of the ALBA project. Peter is an active member of the VSIA Functional Verification DWG. He is currently Director of ASIC Development at Vixel Corporation.

Leena Singh has over nine years experience in ASIC design and verification for multimedia, wireless, and process control applications. Leena graduated with a BSEE from Punjab University, Chandigarh, in India. She was a member of the methodology development for SOC verification at Cadence. She is currently Principal Design Engineer, of Cadence's Vertical Markets and Design Environments.

Acknowledgements

We would like to thank Merrill Hunt, Grant Martin, Dar-Sun Tsien, Larry Drenan, Colette Askeland and Kevin Moynihan for their key support, enthusiastic encouragement and helpful suggestions.

We would like to thank Kolar Kodandapani, Venkatakrishnan Chandran, Larry Cooke, Raminderpal Singh, Narasimha Murthy, Christopher Lennard, Henry Chang, Steve Cox and all the staff members of the Vertical Markets and Design Environments Group at Cadence Design Systems for their enthusiastic encouragement, valuable contributions and support.

We would like to thank Linda Fogel, our editor, for bringing in clarity, consistency and completeness throughout the book.

We would also like to thank Liza Gabrielson, Kerry Burton and Beth Martinez for their creative ideas, hardwork and support in getting this book to production.

In addition, each author would like to especially acknowledge the following members of their families.

Prakash Rashinkar would like to acknowledge his wife, Latha, and son, Pramit, for their whole-hearted cooperation, valuable suggestions and support during the creation of the book.

Peter Paterson would like to acknowledge his wife, Elizabeth, and daughters, Claire and Angela, for their support and understanding during the writing of this book.

Leena Singh would like to express her gratitude to her father, Surinder Bedi, husband, Sarabjeet Singh, and daughter, Keerat, for their love, patience, and unconditional support while writing this book.

The authors would also like to thank all their friends working in various design and services companies who reviewed the chapters and provided valuable feedback. They have all lived up to the popular saying "A friend in need is a friend indeed" by gifting their valuable time to review this book in spite of being heavily busy in this busy-bee Silicon Valley.

"Knowledge is power and there is a happiness in sharing the knowledge amongst a wide population." This is the objective for presenting this book in the SOC era. If you get some useful information and ideas from this book, we feel satisfied with our efforts in creating this book for the design verification community.

Prakash Rashinkar

Peter Paterson

Leena Singh

San Jose, California

Foreword

The major challenge the semiconductor industry is confronted with for the last few years has been to design "system-chips" (or SOCs) with far more complex functionality and domain diversity than in the past, yet in significantly less time. At the very top of the list of challenges to be solved is verification. General agreement among many observers is that verification consumes at least 70 percent of the design effort.

Verifying final design correctness is viewed as the key barrier to designing ever more complex SOCs and exploiting leading-edge process technologies. When the Virtual Socket Interface Alliance (VSIA) held a Verification Workshop in 1999, the conclusion of many of the world's verification experts was "verification is hard." A few weeks later, after considerable in-depth discussion, the final conclusion was "verification is not hard, it is *very* hard." An additional observation was that no single design tool could be used to solve the problem. Instead, a complex sequence of tools and techniques, including classical simulation, directed and random verification, and formal techniques, are needed to reduce the number of design errors to an acceptable minimum.

The third VSIA verification meeting concluded "verification is not just very hard, it is very, very hard." This anecdote motivates this book on SOC verification. Effective verification is fundamental to design reuse, and the gains in productivity that design reuse permits is essential to exploit advanced process technologies. Every reusable design block needs to be accompanied by a reusable "complete" testbench or verification suite for thorough intra-block verification. Furthermore, design teams must learn the discipline of "reuse without rework," that is, to incorporate design blocks and their associated verification suites into SOC designs without modification.

The work of Rashinkar, Paterson, and Singh is a comprehensive guide to an overall SOC verification methodology as well as a description of the arsenal of tools, technologies, and methods available to verification and design engineers. It provides a snapshot of today's verification landscape and broadly outlines the safe pathways through the wilderness, avoiding the swamps and quicksand that lie waiting for the unwary.

The authors treat verification in a logical flow, moving from the system level through individual block verification, both digital and analog/mixed-signal, followed by treatments on simulation, hardware/software co-verification, static netlist verification, and physical verification technologies. Particular attention is paid to newer techniques, including system-level design approaches, testbench migration, formal model and equivalence checking, linting and code coverage, directed random testing, transaction-based verification techniques for testbenches, and a variety of prototyping and emulation approaches. Pragmatic applications of the various techniques and methods are illustrated with an architecture supporting a Bluetooth consumer application. Design files, scripts, verification properties, and configuration files further clarify the techniques to the reader.

Readers will find that this book provides valuable assistance in developing suitable verification strategies for their own SOC and complex ASIC design projects. Using these most advanced techniques will help the industry improve the supporting tools, leading us more quickly toward the dream of design and verification reuse without rework. We applaud this significant contribution.

Grant Martin
Fellow, System Level Design and Verification
Cadence Design Systems, Inc.

Larry Rosenberg
Chair, Technical Committee
Virtual Socket Interface Alliance (VSIA)

San Jose, California

CHAPTER 1

Introduction

The design capacity that can be realized on a single integrated circuit (IC) at deep sub-micron (DSM) technology levels makes it feasible to integrate all major functions of an end product in a single system-on-a-chip (SOC). But the evolution to SOC design presents challenges to the traditional verification approaches.

This chapter addresses the following topics:

- Technology challenges
- Verification technology options
- Verification methodology
- Testbench creation and migration
- Verification languages
- Verification IP reuse
- Verification approaches
- Verification and device test
- Verification plans
- Example of a reference design

The Bluetooth reference design introduced in this chapter is used as an example of the SOC verification methodology described in various chapters of this book.

1.1 Technology Challenges

Silicon technology foundries continue to aggressively shrink the physical dimensions of silicon structures that can be realized on an IC. This shrinkage is accompanied by significant improvements in both circuit capacity and performance. This technology evolution has been characterized by Moore's Law, which states that the ability to integrate logic gates (transistors) onto a single silicon chip doubles every 18 months. Figure 1-1 shows how the current silicon chip technology is evolving.

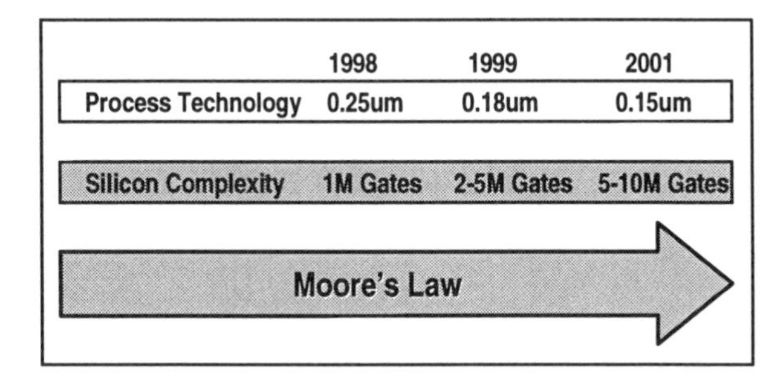

Figure 1-1. Technology Evolution

As silicon technology achieves DSM levels of 0.25 um and below, the design community is confronted with several significant challenges. These challenges can be broadly grouped into three categories:

- Timing closure
- Capacity
- Physical properties

1.1.1 Timing Closure

Traditional silicon design flows have used statistical wire-load models to estimate metal interconnects for pre-layout timing analysis. With this approach, the load on a specific node is estimated by the sum of the input capacitance of the gates being driven and a statistical wire estimate based on the size of the block and the number of gates being driven. For geometries of 0.25 um and above, this approach, which met the timing goals and constraints at the pre-layout stage, could usually be implemented to achieve these same results after physical design.

This success in achieving timing closure was due, in large part, to the fact that, at these geometries, the gate propagation delays and gate load capacitances dominate the total delays and are less sensitive to the interconnect delays. At DSM levels of technology, the interconnect delays become increasingly significant and must be accurately estimated if timing closure is to be achieved. Statistical wire-load models are inaccurate because they represent a statistical value based on the block size. The distribution of wire loads at the mean value can vary greatly, so that the interconnects in the "tail" of the distribution are significantly underestimated.

1.1.2 Capacity

With DSM technology, it is feasible to integrate 10M+ gates onto a single IC using 0.15 um and below technology, which introduces significant capacity challenges to many of the tools in the design flow. To manage this level of complexity, DSM design systems must adopt the following solutions:

- Hierarchical design
- Design reuse

Hierarchical design flows support multiple levels within the design. The top level is an interconnect of the blocks, and the next level down provides the design detail for these blocks, either in terms of interconnected sub-blocks or library elements. By partitioning the design in this way, the complexity of the design, as viewed at a specific level, is constrained. To support this concept, it must be possible to generate abstract models of the blocks at each level of the design for use at the higher levels.

Design reuse integrates preexisting blocks with newly authored blocks. This aides the development of DSM designs in two ways. First, since one or more of the blocks within the design have been pre-designed, the amount of original design work is reduced. Secondly, since the pre-designed blocks have been pre-certified or validated, they can be viewed as black boxes and need not be revalidated.

1.1.3 Physical Properties

At DSM levels of technology, several physical effects need to be accounted for within the design flow. The evolution to DSM results in finer device geometries, more layers of metal interconnect, lower supply voltages, tens of millions of devices within a single IC, lower device thresholds, and higher clock frequencies. These factors cause signal integrity (SI) issues and design integrity (DI) issues to be of greater concern than at more relaxed geometries. SI issues include crosstalk, IR drop, and power and ground bounce. DI issues include electron migration, hot electron and wire self-heating. At more relaxed geometries, these SI and DI issues were lower-order effects and could be ignored.

As geometries have shrunk, these issues become more prevalent so that a sign-off screen is required to check for any violations prior to release for fabrication and, if violations are detected, a fix process is exercised. At DSM levels, these issues are raised to the point where the design process itself must detect and fix violations. Without this "find and fix" capability within the design flow, ten of thousands of violations could occur making it impossible to fix these as a post-processing step. The capacity and physical property challenges of DSM design must be addressed in any SOC verification method.

1.1.4 Design Productivity Gap

Design productivity lags the design density improvements made possible by the technological evolution. Figure 1-2 shows this design productivity gap. The gate density is shown in Gates/chip and the design productivity is shown in Gates/hour.

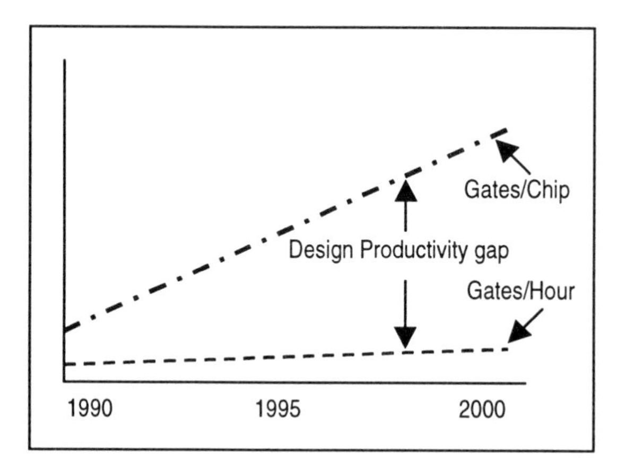

Figure 1-2. Design Productivity Gap

The increasing complexity of ICs poses challenges to both system design engineers and verification engineers. This productivity gap cannot be addressed by simply throwing more engineers at the problem. For one thing, there are not enough qualified engineers to solve the problem and, even if there were, there are practical limitations on how large a design team can grow. As design teams grow, so does the level of coordination required to keep everyone in sync. Instead, new methodologies that make the design process more productive are required.

The industry has responded to this challenge by adopting design reuse strategies. By utilizing preexisting blocks (also know as intellectual property (IP) blocks or virtual components (VC)), the amount of original design work required to realize a new design is reduced. With platform-based design, design reuse goes beyond reusing individual blocks. In platform-based design, a set of core elements that are common across a family of products is identified, integrated, and verified as a single entity. The actual products are then realized by adding individual design elements to this core. The individual elements can be either additional IP blocks or newly authored elements. This concept of design reuse not only reduces the design effort, but also significantly reduces the verification effort in realizing a new design.

1.1.5 Time-to-Market Trends

In conjunction with the explosion in design complexity, dramatic reductions in the time-to-market (TTM) demands for electronic devices are occurring. Figure 1-3 shows the TTM trends for military, industrial, and consumer devices.

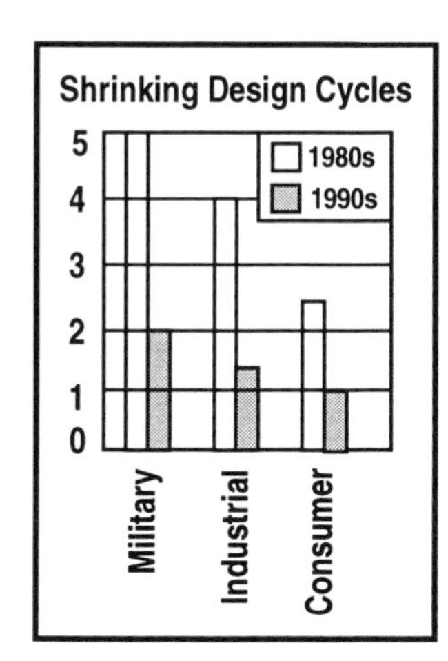

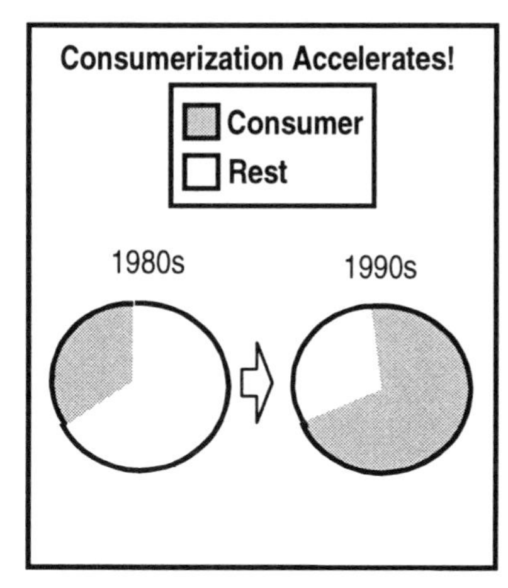

Figure 1-3. Time-to-Market Trends

Not only are the TTM design cycles shrinking for all application spaces, but this is accompanied by a migration away from traditional applications (military and industrial) to consumer products, which have the shortest cycle times. This shift is causing the average cycle time to be shrinking at a faster rate than any one market segment.

These technology and market challenges are having a dramatic impact on verification methodologies and tools. It is estimated that between 40 to 70 percent of the total development effort is consumed by verification tasks. Clearly, these verification activities have to be performed more efficiently if the overall market challenges are to be met.

1.1.6 SOC Technology

The evolution to SOC design has also brought with it challenges to traditional verification approaches. An SOC includes programmable elements (control processors and digital signal processors (DSP)), hardware elements (digital and analog/mixed signal (AMS) blocks), software elements, complex bus architectures, clock and power distribution, test structures, and buses.

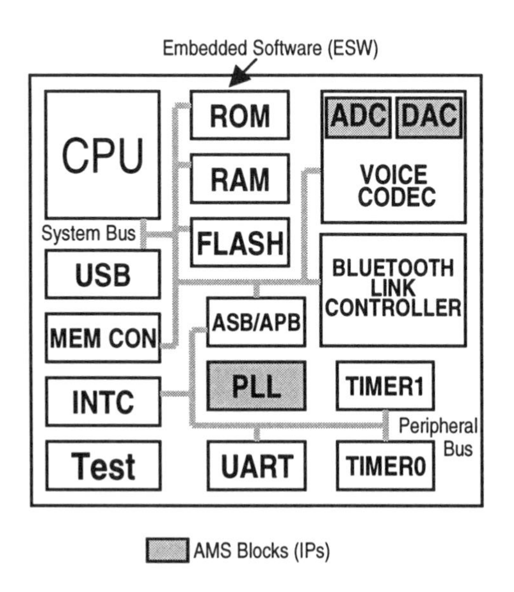

Figure 1-4. System-On-a-Chip

This represents challenges over traditional design methodologies where different design disciplines (digital, AMS, embedded software (ESW)) could be designed in isolation with methodologies and tools specific to each discipline. In SOC, we now have these design disciplines coexisting within a single design, so the verification methodology has to deal with mixed digital and analog verification and mixed hardware/ESW verification. The tools must also deal with the added complexity of SOC devices, not only due to the increased gate counts but also the complex structures and algorithms implemented on these devices.

Figure 1-4 shows a topographical representation of the Bluetooth design to be used as an example throughout this book. In the example Bluetooth design, the radio frequency (RF) portion of the system resides off-chip. This is typical of today's high-speed communications technologies where substrate coupling issues make it impractical to mix RF and baseband circuits. Several emerging technologies, such as silicon-on-insulator and Silicon Germanium, offer the possibility of combining RF and baseband circuits in a single IC.

1.2 Verification Technology Options

The goal of verification is to ensure that the design meets the functional requirements as defined in the functional specification. Verification of SOC devices takes anywhere from 40 to 70 percent of the total development effort for the design. Some of the issues that arise are how much verification is enough, what strategies and technology options to use for verification, and how to plan for and minimize verification time. These issues challenge both verification engineers and verification solution providers.

A wide variety of verification technology options are available within the industry. These options can be broadly categorized into four classifications: simulation-based technologies, static technologies, formal technologies, and physical verification and analysis. The verification technology options currently available are described below. To achieve the SOC verification goals, a combination of these methods must be used.

1.2.1 Simulation Technologies

Simulation technologies include event-based and cycle-based simulators, transaction-based verification, code coverage, AMS simulation, HW/SW co-verification, accelerators, such as emulation, rapid prototype systems, hardware modelers, and hardware accelerators.

1.2.1.1 Event-based Simulators

Event-based simulators operate by taking events, one at a time, and propagating them through a design until a steady state condition is achieved. Any change in input stimulus is identified as an event.The design models include timing and functionality. A design element may be evaluated several times in a single cycle because the different arrival times of the inputs and the feedback of signals from downstream design elements. While this provides a highly accurate simulation environment, the speed of the execution depends on the size of the design and the level of activity within the simulation. For large designs, this can be slow.

Features: Provides an accurate simulation environment that is timing-accurate, and it is easy to detect glitches in the design.

Limitations: The speed depends on the size of the design and the level of activity within the simulation. If the designs are large, the simulation speed may be slow. The speed is limited because event-based simulators use complex algorithms to schedule events, and they evaluate the outputs multiple times.

1.2.1.2 Cycle-based Simulators

Cycle-based simulators have no notion of time within a clock cycle. They evaluate the logic between state elements and/or ports in the single shot. Because each logic element is evaluated only once per cycle, this can significantly increase the speed of execution, but this can lead to simulation errors. Cycle-based simulators only function on synchronous logic.

Features: Provides speeds of 5x to 100x times that of event-based simulators. The simulation speed can be up to 1000 cycles per second for large designs. Best suited for designs requiring large simulation vectors, such as microprocessors, application-specific integrated chips (ASIC), and SOCs.

Limitations: Cannot detect glitches in the design, since they respond only to the clock signal. Also they do not take the timing of the design into consideration, therefore, timing verification needs to be performed using a static-timing analysis tool.

1.2.1.3 Transaction-based Verification

Transaction-based verification allows simulation and debug of the design at the transaction level, in addition to the signal/pin level. All possible transaction types between blocks in a system are created and systematically tested. Transaction-based verification does not require detailed testbenches with large vectors.

The bus function model (BFM) is used in transaction-based verification. BFMs provide a means of running the transactions on the hardware design interfaces. They drive signals on the interconnects according to the requirements of the interface protocols. They can be easily authored in standard hardware description languages (HDL) and C++.

Features: Enhances the verification productivity by rasing the level of abstraction to transaction level, instead of the signal/pin level. Self-checking and directed random testing can be easily performed.

1.2.1.4 Code Coverage

Code coverage analysis provides the capability to quantify the functional coverage that a particular test suite achieves when applied to a specific design. This can be at the individual block level or the full-chip level. The analysis tools provide a value for the percentage coverage of each attribute being assessed, and a list of untested or partially tested areas of the design.

Code coverage analysis is performed on the register-transfer level (RTL) views of the design. It assesses the various types of coverage including: statement, toggle, finite-state-machine (FSM) arc, visited state, triggering, branch, expression, path, and signal.

Features: Provides an assessment of the quality of the test suites. It also identifies untested areas of a design.

1.2.1.5 HW/SW Co-verification

In HW/SW co-verification, integration and verification of the hardware and software occurs concurrently. The co-verification environment provides a graphical user interface (GUI) that is consistent with the current hardware simulators and software emulators/debuggers that are used by the hardware and software project development teams. This enables the software team to execute the software directly on the hardware design. Also, the hardware design is stimulated with real input stimulus, thereby reducing the efforts required to author the hardware testbenches.

Features: Verifies both hardware and software early in the design cycle, enabling fast TTM. It offers sufficient performance to run the interface confidence tests, code segments, and individual driver and utility code.

Limitations: Co-verification environments available today do not offer sufficient performance to run complete application software on top of the target real-time operating system (RTOS) because of capacity and simulation speed problems.

1.2.1.6 Emulation Systems

Emulation systems are specially designed hardware and software systems that typically contain reconfigurable logic, often field programmable gate arrays (FPGA). Some of the emulation systems available in the industry contain high-speed array processors. These systems are programmed to take on the behavior of the target design and can emulate its functionality to the degree that it can be directly connected to the system environment in which the final design is intended to operate. Because these systems are realized in hardware, they can perform at speeds that are orders of magnitude faster than software simulators and, in some instances, can approach the target design speeds.

1.2.1.7 Rapid Prototyping Systems

Rapid prototyping systems can be used to accurately model the prototype of the intended SOC. They are hardware design representations of the design being verified. The key to successful rapid prototyping is to quickly realize the prototype. Some approaches include emulation and reconfigurable prototyping systems, in which the target design is mapped to off-the-shelf devices, such as control processors, DSPs, bonded-out cores, and FPGAs. These components are mounted on daughter boards, which plug into a system interconnect motherboard containing custom programmable interconnect devices that model the target system interconnect.

Application-specific prototypes map the target design to commercially available components and have limited expansion and reuse capability. Typically, these prototypes are built around board support packages (BSP) for the embedded processors, with additional components (memories, FPGAs, and cores) added as needed.

Features: Offers the ability to develop and debug software, giving a real view of SOC hardware. This enables seamless integration of hardware and software when the chip prototype is available. Provides significantly higher simulation speed than software simulators and co-verification.

1.2.1.8 Hardware Accelerators

Hardware acceleration maps some or all of the components in a software simulation into a hardware platform specifically designed to speed up certain simulation operations. Most commonly, the testbench remains running in software, while the actual design being verified is run in the hardware accelerator. Some of the options provide acceleration capability even for testbench.

1.2.1.9 AMS Simulation

The current AMS SOC chip designs are more of top-down digital and bottom-up analog. Analog tools available in the industry provide less automation due to the complex nature of the analog designs.

AMS simulation is more complex than either analog-only or digital-only simulation. The general simulation technique followed is to verify the AMS block independently. The interface part of the AMS block is treated as digital, and the interface is verified after integration with the SOC.

1.2.2 Static Technologies

The static verification technologies include lint checking and static timing verification. This technology does not require testbench or test vectors for carrying out the verification.

1.2.2.1 Lint Checking

Lint checking performs a static check of the design code to verify syntactical correctness. The types of errors uncovered include uninitialized variables, unsupported constructs, and port mismatches. Lint checking can be performed early in the design cycle. It identifies simple errors in the design code that would be time-consuming to uncover with more advanced tools.

1.2.2.2 Static Timing Verification

Each storage element and latch in a design have timing requirements, such as setup, hold, and various delay timings. Timing verification determines whether the timing requirements are being met. Timing verification is challenging for a complex design, since each input can have multiple sources, and the timing can vary depending on the circuit operating condition.

1.2.3 Formal Technologies

In design verification, it is often very difficult to detect bugs that depend on specific sequences of events. These bugs can have a serious impact on the design cycle when they are not detected early in the verification phase. Detecting obscure bugs early on and the exhaustive nature of formal verification have been the main driving forces toward using formal verification techniques. Formal verification methods do not require testbenches or vectors for verification. They theoretically promise a very fast verification time and 100 percent coverage. The formal verification methods are:

- Theorem proving technique
- Formal model checking
- Formal equivalence checking

1.2.3.1 Theorem Proving Technique

The theorem proving technique is still under academic research. This technique shows that the design meets the functional requirements by allowing the user to construct a proof of the design behavior using theorems.

1.2.3.2 Formal Model Checking

Formal model checking exploits formal mathematical techniques to verify behavioral properties of a design. A model checking tool compares the design behavior to a set of logical properties defined by the user. The properties defined are directly extracted from the design specifications. Formal model checking is well suited to complex control structures, such as bus arbiters, decoders, processor-to-peripheral bridges, and so on.

Features: Formal model checking tools do not require any testbenches or vectors. The properties to be verified are specified in the form of queries. When the tool finds an error, it generates a complete trace from initial state to the state where the specified property failed.

Limitations: Formal model checking does not eliminate the need for simulation but rather supplements it. These tools do not consider timing information of the design; timing verification needs be done using a timing analysis tool. Current model checkers have capacity restrictions limiting their usage to small designs. If the design properties are not represented correctly, errors might not be detected.

1.2.3.3 Formal Equivalence Checking

Formal equivalence checking is a method of proving the equivalence of two different views of the same logic design. It uses mathematical techniques to verify equivalence of a reference design and a modified design. These tools can be used to verify the equivalence of RTL-RTL, RTL-Gate, and Gate-Gate implementations. Since equivalence checking tools compare the target design with the reference design, it is critical that the reference design is functionally correct.

Features: Faster than performing an exhaustive simulation. It is guaranteed to provide a 100 percent verification coverage and does not require developing testbenches and vectors.

Limitations: Does not verify the timing of the design, so it must be used in conjunction with a timing analysis tool.

1.2.4 Physical Verification and Analysis

In DSM designs, all electrical issues and processes must be considered to understand and fix the effects of interconnect parasitics, since interconnect delays dominate the gate delays. The issues that are to be analyzed and fixed are timing, signal integrity, crosstalk, IR drop, electromigration, power analysis, process antenna effects, phase shift mask, and optical proximity correction. Options that perform estimation during pre-layout phase and extraction and analysis of physical effects after post-layout phase are available in the industry.

1.2.5 Comparing Verification Options

The various verification technology options available in the industry offer different advantages and features. Table 1-1 compares what the different verification techniques offer. In addition, which option to use depends on the design complexity, the verification requirements, and what can be afforded.

Table 1-1. Comparing Verification Options

	Event-based Simulation	Cycle-based Simulation	Hardware Accelerators	Emulation	Formal Verification	Static Timing Verification
Function	Yes	Yes	Yes	Yes	No	No
Abstraction Level	Behavioral, RTL, Gate	RTL, Gate	RTL, Gate	RTL, Gate	RTL, Gate	Gate
Functional Equivalence	Yes	Yes	Yes	Yes	Yes	No
Timing	Yes	No	Yes/No	No	No	Yes
Gate Capacity	Low	Medium	High	Very high	High	Medium
Run Time	<10 Cycles	1K Cycles	1K Cycles	1M Cycles	Medium	High
Cost	Low	Medium	Medium	High	Medium	Low

Table 1-2 compares the various technology options for HW/SW co-verification. The ISS option is for the processor, and the C models are for peripherals, HW/SW co-verification tools, rapid prototype. The features compared are what system speed is achievable, system or chip timing, the ability to run application software, the availability of a debugging facility, and the overall cost.

Table 1-2. Comparing HW/SW Co-verification Options

Option	**Speed**	**Timing**	**Software**	**Debug**	**Cost**
Instruction set simulation (ISS)	Medium	No	C Model	Algorithm	Low
HW/SW co-verification	Slow	Yes	Real	HW/SW	High
Rapid prototype	Fast	Yes	Real	Low	Medium
Emulation	Very fast	Yes	Real	Low	Very high

1.2.5.1 Which Is the Fastest Option

Figure 1-5 shows the growing gap between the increasing demand for verification and the simulation technology performance offered by the various options.

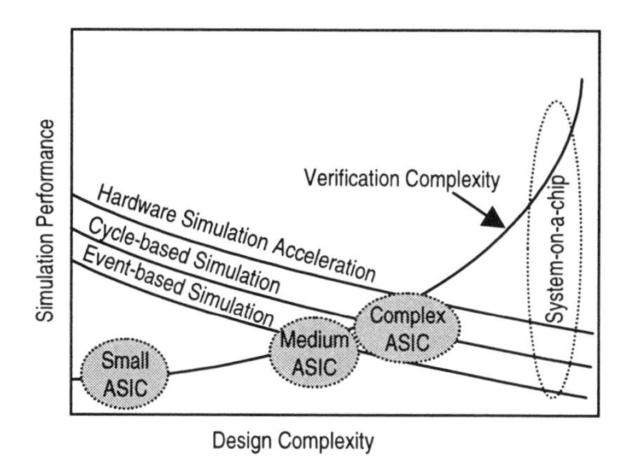

Figure 1-5. Growing Gap

When determining which solution to select for verifying a design under test (DUT), the obvious question that arises is which is the fastest method. Since each verification technique has its advantages and disadvantages, the answer is not straight-forward. The following lists the appropriate application scenarios for the various methods.

- **Event-based simulation**: Best suited for asynchronous designs. It focusses on both function and timing of the design. It is also the fastest technique for small designs.
- **Cycle-based simulation**: Focusses only on the function of the design; no timing. It is useful for medium-sized designs.
- **Formal verification**: Does not require test vectors or testbenches. It includes model checking and equivalence checking. The model checking verifies the control logic, and it requires that the properties and constraints of the design be defined. The current model-checking tools lack the design capacity, so they are useful for smaller designs only. Equivalence checking is used between two versions of the same design. It can handle designs with larger capacity.
- **Emulation**: Performs the verification of the design at much higher simulation speeds than other techniques. It can handle very large capacity designs and test vectors.

- **Rapid prototype**: Useful for developing software for the product. Provides early access to the SOC platform, enabling the engineers to develop the software and test, thereby improving overall TTM goals.

Except for event-based simulation, these methods check the functionality of the design only. The timing must be checked using static timing analysis tools.

1.3 Verification Methodology

Design verification planning should start concurrent with the creation of specifications for the system. System specifications drive the verification strategy. Figure 1-6 shows the high-level verification flow for an SOC device. This section focusses mainly on verification methodology aspects of SOC.

1.3.1 System-Level Verification

In system design, system behavior is modeled according to the specifications. The system behavior is verified using a behavioral simulation testbench. The behavioral testbench might be created in HDL, C, C++, or with a testbench language such as Vera or Specman Elite.

Once the system behavior is validated, the system is mapped to a suitable architecture using hardware and software IPs available in the library or authored as part of the design process. The hardware and software partitioning is done. The function and performance of the architecture is verified with the testbench that is created during the system behavioral simulation.

The testbench created for system-level verification is generally token-based and does not address the cycle-accurate and pin-accurate aspects of the system. The testbench should be converted to a suitable format, so it can be used for hardware RTL code simulation and software verification.

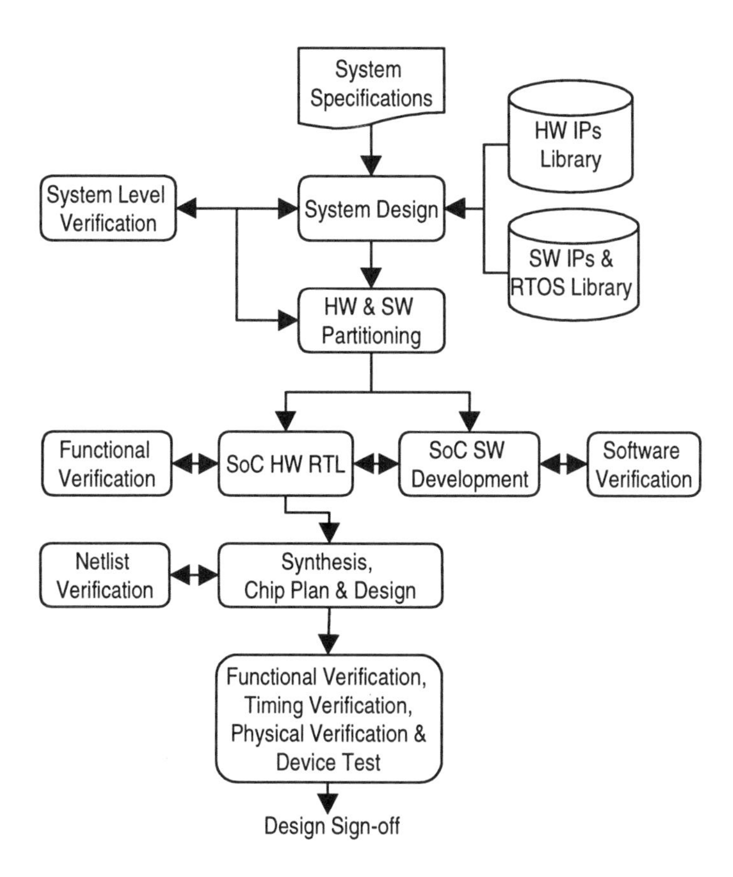

Figure 1-6. SOC Verification Methodology

1.3.2 SOC Hardware RTL Verification

In hardware verification, the RTL code and testbench are obtained from the system design. The testbench is converted or migrated to a suitable format to verify the RTL code and the design is verified for functionality. The verification mainly focusses on the functional aspects of the design. RTL verification includes lint checking, formal model checking, logic simulation, transaction-based verification, and code coverage analysis.

1.3.3 SOC Software Verification

In software verification, the software and test files are obtained from the software team. Software verification is performed against the specifications obtained from the system design. Depending on the verification requirements, verification and hardware/software integration can be performed using soft prototype, rapid prototype system, emulation, or HW/SW co-verification.

1.3.4 Netlist Verification

The hardware RTL is synthesized and a gate-level netlist is generated. The netlist can be verified using a formal equivalence checking tool with the RTL code as the reference design and the gate-level netlist as the implementation design. This ensures that the RTL and gate-level netlist are logically equivalent.

The netlist undergoes changes when the clock tree and scan chain are added to the design. The design is verified using a formal equivalence checking tool after clock-tree generation and scan-chain insertion to ensure the correctness of the design.

Timing verification is carried out during various steps of the chip plan/design phase to ensure that the design meets the timing requirements.

1.3.5 Physical Verification

Physical verification is performed on the chip design to ensure that there are no physical violations in the implemented design. The physical verification includes design rules checking, layout versus schematic, process antenna effects analysis, SI checking, including crosstalk, and current-resistance (IR) drop.

1.3.6 Device Test

The final device test uses the test vectors that are generated during the functional verification. The device test checks whether the device was manufactured correctly. The device test focuses on the structure of the chip, such as wire connections and the gate truth tables, rather than chip functionality.

Vectors are generated for manufacturing the device test using the testbench created during functional verification and/or using an automatic test pattern generator (ATPG) tool. After the verification produces satisfactory results, the design is ready for fabrication sign-off and tape-out.

1.4 Testbench Creation

A testbench is a layer of code that is created to apply input patterns (stimulus) to the design under test (DUT) and to determine whether the DUT produces the outputs expected. Figure 1-7 shows a simple block diagram of a testbench that surrounds the DUT.

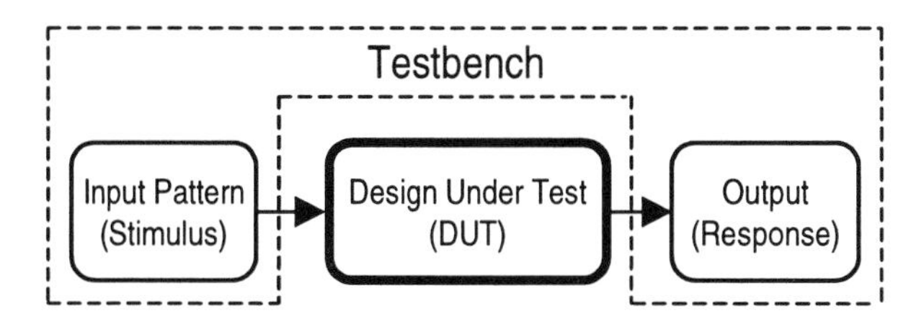

Figure 1-7. Testbench

A testbench that is created to apply inputs, sample the outputs of the DUT, and compare the outputs with the expected (golden) results is called a self-checking testbench, as shown in Figure 1-8. A self-checking testbench generates errors if the sampled outputs do not match the expected results. It also indicates the status of the inputs, outputs obtained, and expected results along with error information. This helps in analyzing the DUT's detail functionality and isolating the cause of the error.

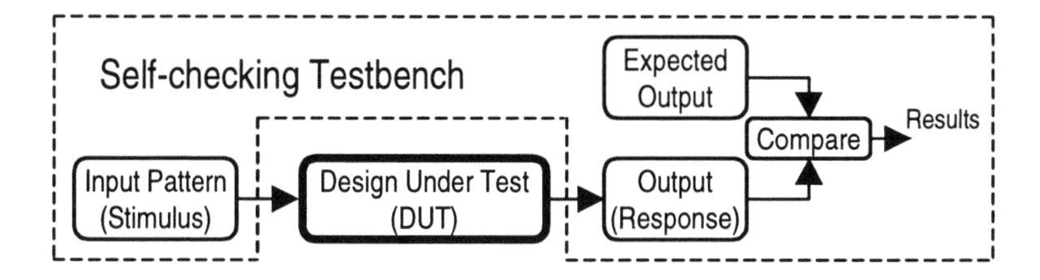

Figure 1-8. Self-checking Testbench

Creating self-checking testbenches for all designs is recommended, since it provides an easy way to detect, understand, and fix errors. The alternative approach of visually checking results is highly error prone. A self-checking testbench written with clear comments in the code helps engineers not involved in the project to understand and start working in the project quickly.

The process of creating a testbench for a design involves thorough understanding of the functional specifications. The following techniques can be used for testbench creation:

- Testbench in HDL
- Testbench in programmable language interface (PLI)
- Waveform-based
- Transaction-based
- Specification-based

1.4.1 Testbench in HDL

The testbench is created using a standard HDL, such as Verilog or VHDL. This technique works fine for small designs. The testbench becomes complex and difficult to maintain if the design verification requirements increase.

1.4.2 Testbench in PLI

The C programming language is used to write a PLI testbench. This can be linked to an HDL simulator. All the tasks defined in PLI can be called from within the procedural block in the HDL code. This technique can be used when the testbench needs to simulate more complex functions.

1.4.3 Waveform-based

In this technique, the signal waveforms are edited according to the design requirements. A tool that converts the waveforms into a stimulus generation embedding the timing information is used to create a testbench. The testbench is then used to verify the DUT.

1.4.4 Transaction-based

This technique uses bus function models (BFMs) and transaction-level stimulus concepts. The testbench blocks are modeled using BFMs that translate the transaction-level test stimulus into cycle-accurate and pin-accurate transitions at the DUT interface. The transactions are performed according to the protocols used in the design. The response from the DUT is translated from the pin- and cycle-accurate response of the DUT back to a transaction-level response. Stimulus and responses can be viewed at the transaction level, in addition to the signal/pin level. Typically, responses are checked at the transaction level and when a mismatch is identified, this can be isolated by further viewing at the signal/pin accurate level. Both the BFMs and the transactions created can be reused with no or few modifications for other similar designs.

1.4.5 Specification-based

In a specification-based technique, the testbench is created by capturing the specification of the design in an executable form. This enhances the productivity, since the abstraction level for verification is increased to specification level rather than RTL. This enables the verification engineers to focus on design intent instead of spending time in creating complex testbenches focussed on design implementation.

1.5 Testbench Migration

The testbench created for system-level verification is generally token-based, which is not suitable for performing verification on lower-level views of the design, such as RTL and gate-level netlist. To test these levels involves migrating the testbench from one level to the next, as follows:

- Translate the stimulus from the upper level to a format suitable for application at the next lower level.
- Apply the testbenches to both levels of the design and compare the results at the points of indifference between the designs.
- Extract a new version of the testbench from the lower-level model containing the additional details.

To facilitate testbench migration from the functional level to lower levels, use the following representations in the functional testbench:

- **Bit-true representations**: Data values in the programming language (C/C++) that the testbench is created in have no concept of bus width. Hardware implementations of these functions have a fixed-bus width. Modeling at the functional level using bit-true representations ensures convergence of results.

- **Fixed-point representations**: Fixed-point implementation is used for functional modeling. This aids in the alignment of the functional model and hardware implementation.

1.5.1 Testbench Migration from Functional to RTL

The system-level functional design is typically written in C or behavioral HDL, and the associated testbench deals with token or frame-based transactions. In this instance, a token is a data block of variable size. The functional model has no concept of time or clocks, and the transaction testbench is applied by an event scheduler.

The results of this testbench are usually written to an external memory, and the success or failure of the test is determined by the memory contents when the test is completed. To migrate this test to an RTL model, the testbench must be translated into pin- and bus-level cycles with the associated clocks. The results are checked by comparing the external memory contents, created by running the functional test on the functional model, to the migrated test run on the RTL.

Once the migrated test run matches the external memory contents at the points of indifference between these models, a new testbench can be created by capturing the cycle-by-cycle behavior of the RTL model. This cycle-by-cycle behavior can be captured at the I/Os of the design or include internal state traces. This new testbench can then be used to compare the RTL to lower design abstractions.

1.5.2 Testbench Migration from RTL to Netlist

To achieve testbench migration from RTL to netlist, the testbench created from the RTL model is transformed into a format suitable for application to the netlist level. The bus-based RTL testbench is translated into a bit- and pin-accurate stimulus. This stimulus can then be applied to the netlist model and the results compared to the RTL response at the points of indifference, which are the I/O pins and internal state elements. These comparison points are sampled at the end of each cycle. Once these have been verified as matching, a more detailed testbench can be created by capturing the switching times within a cycle for the output transitions.

1.6 Verification Languages

The Specman Elite and Vera hardware verification languages offer enhanced productivity to verification engineers by reducing the problems and time spent in creating complex testbenches. Design teams using these verification languages report productivity improvements of a factor of 4x times. The testbenches created are compact and easy to understand.

Specman Elite generates tests automatically by capturing the rules from the design specification. The automation includes generating functional tests, checking verification results, and coverage analysis. Specman Elite also provides ease of verification testbench reuse.

The Vera hardware verification language creates target model environments, automatic stimulus generation, self-checking tests with error indication, and coverage analysis capabilities. Vera also provides features for easy verification testbench reuse.

Legacy testbenches created in Verilog, VHDL, and C/C++ can be used with testbenches generated using Specman Elite or Vera, thus protecting investments already made. These languages interface with HW/SW co-verification tools.

Testbenches can also be created in Verilog, VHDL, PLI, or C++ languages. Cadence's TestBuilder tool, which is part of Cadence® Verification Cockpit, a comprehensive library of object-oriented techniques, can be used to create testbenches in C++.

1.7 Verification IP Reuse

TTM pressures on product design cycles are forcing SOC designers and integrators to reuse available design blocks. To ensure that designs are reusable, the design teams use guidelines, such as modularity with good coding style, clear comments in the code, good documentation, and verification testbenches.

Just as the HDL-based design blocks are reusable, it is desirable to reuse the verification testbench's IP to enhance overall productivity. Verification teams put a lot of effort into developing testbenches. If the testbenches developed for one design can be used for other similar designs, a significant amount of the verification time is saved for subsequent designs. For example, the testbench developed to verify a platform-based SOC for multimedia audio applications can be used for a multimedia voice-over-packet application with minor modifications to the platform SOC. If the system is verified using transaction-based verification, the transactors and test suites developed for the platform can be reused for the voice-over-packet application.

It is important to create standard stimulus formats so that the patterns used for functional verification can be reused across multiple projects. For example, sample global system for mobile communications (GSM) and automatic transfer mode (ATM) packet data can be created for use in multiple projects. Similarly, testbenches developed for verifying advanced microcontroller bus architecture (AMBA) or peripheral component interconnect (PCI) bus protocols or functionality can be reused for any system based on these buses.

Modular and reusable testbenches can be developed with TestBuilder, Vera, and Specman Elite.

1.8 Verification Approaches

SOC design houses use various verification approaches. These include top-down design and verification, bottom-up verification, platform-based verification, and system interface-driven verification.

1.8.1 Top-Down Design and Verification Approach

Figure 1-9 shows the methodology for top-down design and verification. The starting point for any top-down design is the functional specification. This can be an executable specification, but typically it is a written specification with all of the associated ambiguities of natural languages. From the functional specification, a detailed verification plan is developed (described later in this chapter).

A system model of the design is developed from the functional specification. It is created by integrating system-level models for the component elements of the design, authored in a system-level language (C, C++, and others).

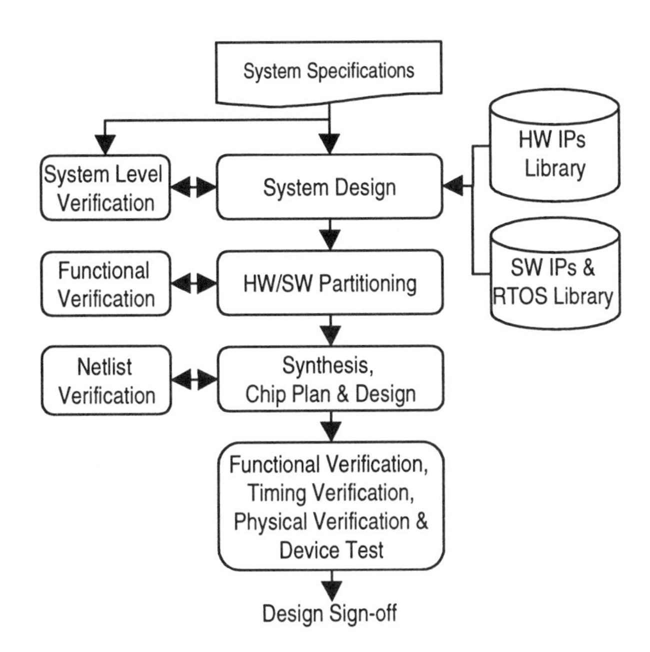

Figure 1-9. Top-Down Design and Verification Approach

The system-level model is functionally verified by exercising it with the system-level testbench. The design can then be decomposed through any number of abstraction levels until the detailed design is complete. Possible abstraction levels include architectural, HDL, and netlist models. At the upper abstraction levels, the design is verified using the system testbench, which is enhanced at each abstraction level to test the increased functional and temporal detail at that level.

Transaction-based verification can be used for system-level verification and interconnect verification. Additional functionality that requires testing as the design progresses include the following:

- Transition from a transaction- or token-based model to a pin-accurate model
- Transition from a token-passing model to a protocol-based communication model
- Inclusion of test structures

Once the design has been refined to the HDL level, additional verification technologies may be applied to test aspects that cannot be verified through simulation. These technologies include lint checking, which verifies that the incoming code conforms to coding standards and guidelines, and formal model checking, which exhaustively checks design properties.

After the design's RTL is verified, either formal equivalence checking tools or simulation is used to verify the implementation views.

After the above tests, timing verification, physical verification, and device tests are performed to ensure correct chip implementation.

As design sizes increase, it might not be feasible to run the full system-level test suite on a detailed model. In this case, some options to consider are accelerating the simulation environment using emulation, rapid prototype system, or hardware accelerators or partitioning the design into several functional blocks. Here, the system-level block test is extracted from an abstract model of the design running the full system testbench. The individual blocks can then be verified in isolation, with their associated system test suite. When partitioning the design, system simulations can run in a mixed-level mode where most blocks are run with abstract models of the design, and the detailed design is substituted for each block in turn.

1.8.2 Bottom-Up Verification Approach

Figure 1-10 shows the bottom-up verification flow for an SOC device. This approach is widely used today in many design houses. The first step is to validate the incoming design data by passing the files through a parser to ensure that they are compatible with the target tools. Where a tool supports a subset of the total language, the incoming design files must be screened to ensure that they only use the permitted subset of functions and constraints.

Next, the design files are passed through a lint checker to verify that no syntactical design violations within the code (uninitialized variables, unsupported constructs, port mismatches, and so on) exist. The next steps depend on the abstraction level of the design. If the design is at the detailed design level, the flow proceeds sequentially through levels 0, 1, 2, and 3 testing. If the design is at a higher level of abstraction, it can proceed directly to system-level testing at level 3. Verification levels are defined as follows:

- **Level 0**: Verifies the individual components, blocks, or units in isolation. The intent is to test the component exhaustively without regard to the environment into which it will be integrated. The techniques and technologies used in unit test are identical to those applicable to an integrated design: deterministic simulation, directed random simulation, lint, and formal checking. The level 0 tests and associated testbench might be supplied by the IP vendor. In this case, level 0 testing might include code coverage analysis to ensure the quality of the design. If the IP block comes from an IP library, then include level 0 testing as part of the incoming acceptance and conformance testing.

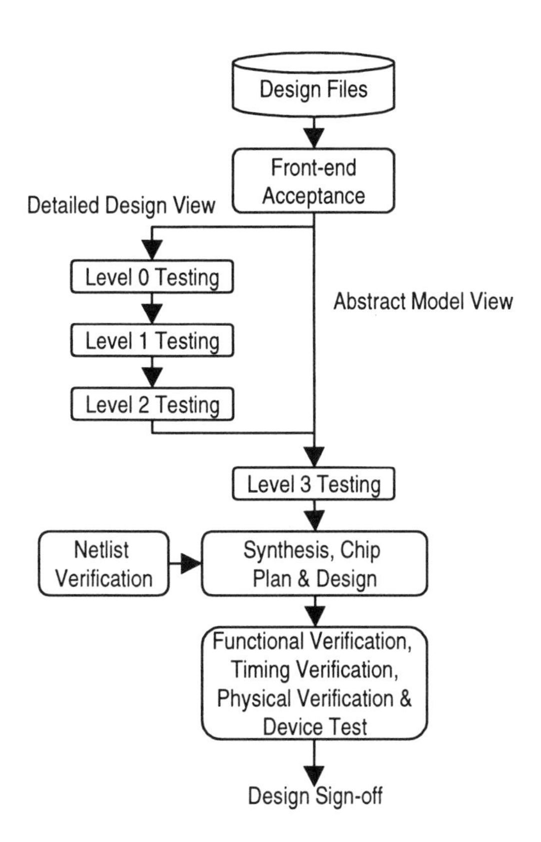

Figure 1-10. Bottom-up Verification Approach

- **Level 1**: Verifies the system memory map and the internal interconnect of the design. These tests can either be created manually or generated automatically by a tool that reads the system-level interconnect and memory map. These tests check that each register in the design can be written to and read by the on-chip processor. In addition, all interconnect within the design is verified by performing writes and read-backs across all the communication paths within the design. These tests typically run in the local memory for the on-chip processor and are self-checking.

- **Level 2**: Verifies the basic functionality of the design and the external interconnect. Tests are written to exercise the main functional paths within each of the functional blocks and to exercise each of the I/O pins.
- **Level 3**: Verifies design at the system level. The goal is to test the functionality of the integrated design exhaustively. Special attention should be paid to corner cases, boundary conditions, design discontinuities, error conditions, and exception handling to ensure a comprehensive test.

After the above tests, the netlist verification, timing verification, physical verification, and device tests are performed to ensure correct chip implementation.

1.8.3 Platform-based Verification Approach

This approach, as shown in Figure 1-11, is suitable for verifying derivative designs based on a preexisting platform that is already verified. It assumes that the basic platform, hardware IP, and software IP used have been verified.

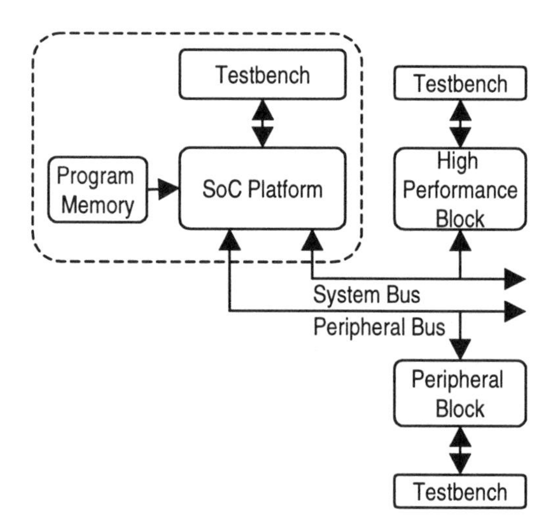

Figure 1-11. Platform-based Verification Approach

The additional IPs are added according to the derivative design requirements. The verification for such a design involves the interconnect verification between the basic platform and the additional IP blocks. The platform itself might be verified using the top-down or bottom-up verification approach.

1.8.4 System Interface-driven Verification Approach

In system interface-driven approach, the blocks to be used in the design are modeled at their interface level during system design. These models, along with the specifications for the blocks to be designed and verified, are handed off to the design team members. The interface models can be used by the verification engineers to verify the interface between the designed block and the system. This eases the final integration efforts and enables early detection of errors in the design.

In the example shown in Figure 1-12, the system consists of five blocks. Block E is being designed and verified using the interface models of blocks A, B, C, and D. If block A is to be verified, then block E is replaced with its interface model, and so on.

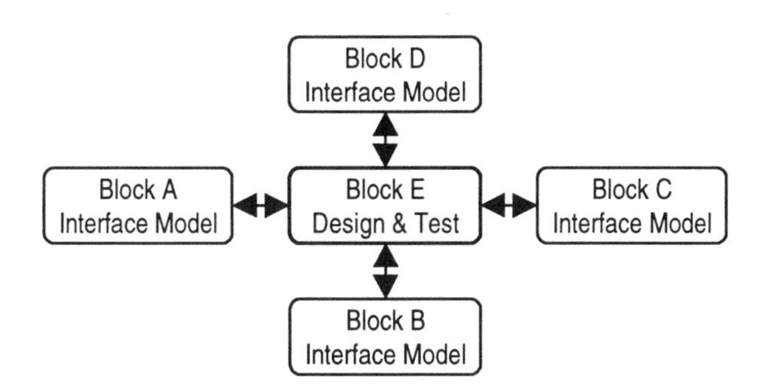

Figure 1-12. System Interface-driven Verification Approach

1.9 Verification and Device Test

Verification of a design is performed to ensure that the behavior, functions, and operation of the design conforms to the functional specification. A device test is performed to check that specific devices have been manufactured defect free. The vectors used for a device test focus on the structure of the chip, such as wire connections and the gate truth tables, rather than chip functionality.

1.9.1 Device Test Challenges

SOC devices are composed of control processors, DSPs, memories, control logic, application specific design cores, and AMS cores. This level of complexity poses the following challenges when selecting and implementing an appropriate device test methodology.

- **Test vectors**: The number of test vectors required for full coverage is enormous. The device test methodologies and test generation/validation tools must be capable of handling the capacity and speed.
- **Core forms**: The cores embedded within the SOC come in different forms, such as soft cores (synthesizable RTL code), firm cores (gate-level netlist), and hard cores (already laid out). Since the core forms are developed by different IP providers, different test techniques might have been used.
- **Cores**: In the past, the functions of logic, memory, and AMS cores functions were implemented in separate chips and manufactured and tested using test methodologies, tools, and automatic test equipment specific to the design style. With SOC, these blocks are embedded in the same SOC, which requires integrated and cost-effective testing.
- **Accessibility**: Accessing and isolating embedded cores is often very difficult and/or expensive.

1.9.2 Test Strategies

The following test strategies are used for various types of cores in an SOC:

- **Logic BIST** (built-in-self-test): Tests logic circuits. The stimulus generators and/or the response verifiers are embedded within the circuit. Generally, this technique employs linear feedback shift registers to generate pseudo-random pattern generation and output pattern signature analysis.
- **Memory BIST**: Tests memory cores. An on-chip address generator, data generator, and read/write controller that applies a common memory test algorithm to test the memory are incorporated.
- **Mixed-signal BIST**: Used for AMS cores, such as ADC, DAC, and PLL.
- **Scan chain**: Assesses timing and structural compliance. A set of scan cells are connected into a shift register by connecting the scan cell's output port of one

flip-flop to the dedicated scan input port of the proceeding scan flip-flop. A scan cell is a sequential element connected as part of a scan chain. Once the scan chains have been inserted into the design, ATPG tools can be used to generate manufacturing tests automatically.

Many companies offer test solutions for SOC. The test methodology should be considered before designing a chip, since it affects the overall design methodology, schedule, and cost.

1.10 Verification Plans

For any SOC design, it is important to develop and document a verification plan to serve as a map for all of the verification activities to be performed. This plan identifies the resources necessary to execute the plan and defines the metrics and metric values that must be achieved for the plan to be accepted as complete.

All plans attempt to predict the future: an imprecise science. The verification team must be prepared to modify the plan if unanticipated problems and limitations arise. If a plan is modified, the changes must still meet the overall goals of the plan.

The verification plan should address the following areas.

1.10.1 Project Functional Overview

The project functional overview summarizes the design functionality, defines all external interfaces, and lists the major blocks to be used.

1.10.2 Verification Approach

Which verification approach fits the design requirements, for example:

- Top-down design and verification
- Bottom-up verification
- Platform-based verification
- System interface-driven verification

1.10.3 Abstraction Levels

The abstraction levels to be verified include behavioral, functional, gate level, and switch level.

1.10.4 Verification Technologies

The verification options or methodologies to be used. For example:

- Dynamic verification tools

  Exercises a model of the design with specific stimulus, checking that the design functions as expected
- Static verification tools

  Confirms that the design exhibits specific behaviors by mathematical analysis of the design. Exhaustively checks the functionality for the state space identified
- Co-verification
- Rapid prototype
- Emulation system
- Static timing verification tools
- Physical verification and analysis tools
- Device test tools

1.10.5 Abstraction Level for Intent Verification

The abstraction level that the design functionality is verified against. The functionality of design abstractions below this level is done by proving their functional equivalence to this model.

1.10.6 Test Application Approach

The approach for applying functional tests to the design. Typically, this involves either pre-loading tests into on-chip memory and executing the test via the on-chip processor or applying the test throughout the external interfaces to the device (test and/or functional interfaces).

1.10.7 Results Checking

How to verify the design's responses to functional tests. This can be done by self-checking techniques, golden model (reference model) comparison, or comparing expected results files.

1.10.8 Functional Verification Flow

Which functional verification tools, databases, and data formats to use when migrating from one tool to another.

1.10.9 Test Definitions

Defines the tests that are to be performed and the model abstraction levels to which the tests are to be applied. In many instances, the same test will be applied to several model levels. For each test, the verification technology or tool and the associated metrics should be included. The metric indicates when the test is complete. Metrics can be defined for combinations of tests. For example, 100 percent statement coverage will be achieved when executing all of the simulation tests.

1.10.10 Testbench Requirements

The testbench requirements based on analyzing the contents of the verification definition table. The model types and abstraction levels, model sources, and testbench elements (checkers, stimulus, and so on) need to be considered. For formal verification, define design properties and constraints.

1.10.11 Models

Which models to use for functional verification. The models include behavioral, functional, RTL, logic level, gate level, switch level, and circuit level.

- **Model sources**: It is important to identify the source for all of the models required in the verification plan and when they will be available.
- **Existing models**: Existing models might be available as part of a standard library, be provided as a supporting model for a core supplied (either directly by the core provider or by a third-party supplier), or be a model that was created for a previous design and is available for reuse. A plan for acquiring the models and making them available to the verification team should be developed.

- **Derived models**: Derived models are created as part of the design flow. For example, if an RTL model of an element is available, a gate-level model can be derived through synthesis. The project plan must show that all derived models will be created before they are required for verification and that the development plan has explicit dependencies showing this requirement.
- **Authored models**: Models to be developed as part of the overall development plan. A plan for developing the models and making them available to the verification team should be determined.

1.10.12 Testbench Elements

The testbench elements include the following.

- **Checkers**: Includes protocol, expected results, and golden model checkers.
- **Transaction verification modules**: A collection of tasks, each of which executes a particular kind of transaction. The module connects to the DUT at a design interface. Because most designs have multiple interfaces, they also have multiple transaction verification modules to drive stimulus and check results.
- **Stimulus**: A set of test vectors based on the design specification (for example, the data sheet). All boundary conditions are covered. The stimulus is created to check whether the model covers exception cases in addition to the regular functional cases. Test vectors are broken down into smaller sets of vectors, each testing a feature or a set of features of the model. These test vectors are built incrementally. Following is a list of vectors.

  -Various data patterns, for example, all ones and zeroes 0xff, 0x00, walking ones and zeroes 0xaa, 0x55

  -Deterministic boundary test vectors, like FIFO full and empty tests

  -Test vectors of asynchronous interactions, including clock/data margining

  -Bus conflicts, for example, bus arbitration tests in full SOC tests

  -Implicit test vectors that might not be mentioned in the data sheet, for example, action taken by a master on a bus retract, system boot, system multitasking and exception handling, and randomly generated full-system tests

1.10.13 Verification Metrics

Two classes of metrics should be addressed in the verification plan:

- **Capacity metrics**: Identifies tool capacity assumptions (run times, memory size, disk size, and so on) and verifies that the assumptions made in developing the verification plan hold true during the execution of that plan.
- **Quality metrics**: Establishes when a verification task is complete. Quality metrics include functional coverage and code coverage.

1.10.14 Regression Testing

The strategy for regression testing. The test plan details when the regression tests are to be run (overnight, continuously, triggered by change levels, and so on) and specifies the resources needed for the regression testing. Typically, once the design has been realized and verified at a certain level, a formal control procedure is put in place to manage the design updates to this golden model and the subsequent re-verification. The test plan should clearly state at what level of abstraction the regression tests are to be run and identify the tests to be used. The regression test suite might be the full set of tests identified for the level of abstraction of the design or a selected subset.

1.10.15 Issue Tracking and Management

Which tracking system to use to manage bugs and errors found in the design.

1.10.16 Resource Plan

The resources required to execute the verification plan, such as human resources, machine resources, and software tool resources.

1.10.17 Project Schedule

The tasks that must be performed to execute the verification plan as well as key benchmarks and completion dates. The plan should show the interdependencies between tasks, resource allocation, and task durations.

1.11 Bluetooth SOC: A Reference Design

In this book, the Bluetooth protocol controller application is used as a reference design to illustrate SOC verification methodology.

Bluetooth is an open protocol standard specification for short-range wireless connectivity. Over 1,200 companies are members of the Bluetooth consortium. Figure 1-13 shows Bluetooth devices.

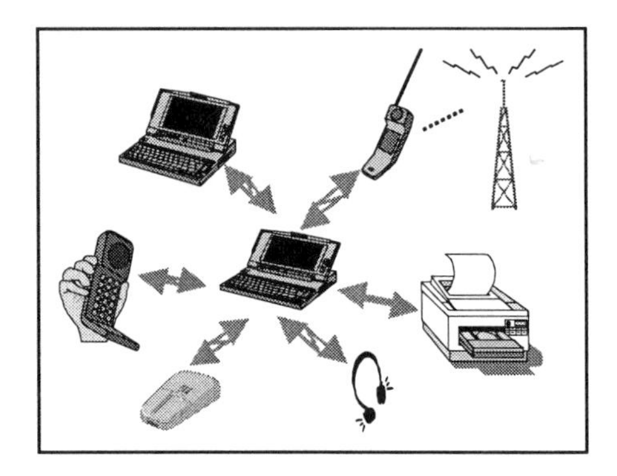

Figure 1-13. Bluetooth Device Network

The Bluetooth protocol offers the following:

- Enables users to make instant connections between a wide range of communication devices
- Complete wireless specification and link with normal modules up to 10 meters or, with a power amplifier, up to 100 meters
- Supports up to seven simultaneous links
- Data rate of 1 Mega symbols per second
- Operates in the Industrial, Scientific, and Medical (ISM) frequency band of 2.4 GHz, ensuring worldwide communication compatibility
- Frequency-hopping radio link for fast and secure voice and data transmission
- Supports point-to-point and point-to-multipoint (including broadcast) connectivity
- Low power consumption

Refer to www.bluetooth.com for more details on the Bluetooth protocol standard.

1.11.1 Bluetooth Device Elements

As shown in Figure 1-14, the Bluetooth device consists of a radio block and a Bluetooth protocol controller block (also referred to as the Bluetooth SOC in this book).

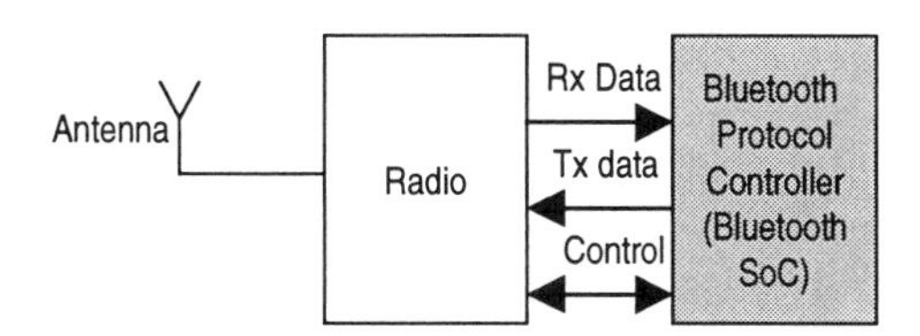

Figure 1-14. Block Diagram of a Bluetooth Device

The radio block communicates with other devices. It interfaces with an antenna on one side, and the Bluetooth protocol controller block on the other side. The Bluetooth SOC handles all the baseband functions, the device control, and the data processing operations. Some chip vendors are offering a single chip solution in which both radio and protocol controller are embedded.

1.11.2 Bluetooth Network

A Bluetooth network consists of master and slave devices, as shown in Figure 1-15. The existence of two or more devices constitutes a piconet. The master owns the piconet and can have up to seven slaves simultaneously. The same unit can be a master or a slave. The device establishing the piconet becomes the master, all other devices will be slaves. Within a piconet, there can be only one master. Each device is given unique address identification and differing clock offsets in the piconet. The slaves can be active or in hold mode. Also, a piconet can have up to 200+ passive slaves. The maximum capacity for each piconet is 1 Mega symbols per second, which is shared among all the slave devices. Ten piconets can be in the same proximity.

Bluetooth uses a 2.4GHz frequency band, with frequency hopping/time-division duplex scheme for transmission. It uses a fast hopping rate of 1,600 hops per second. The time interval between two hops is 625 microseconds and is called as a slot. Each slot uses a different frequency. Bluetooth uses 79 hop carriers equally spaced with 1 MHz. A piconet consists of all devices using the same hopping sequence with the same phase.

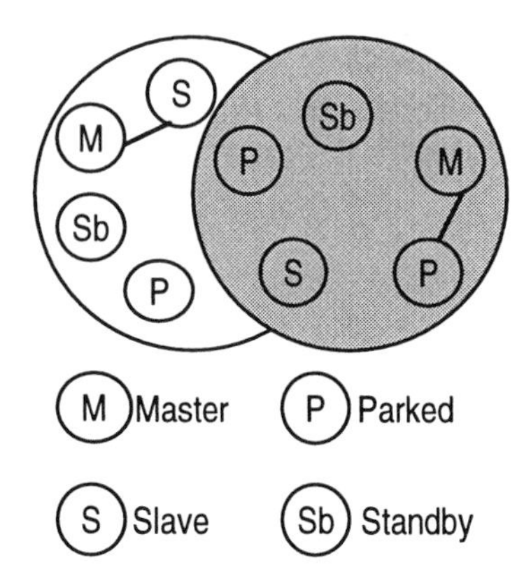

Figure 1-15. Bluetooth Piconet Example

As shown in Figure 1-16, the Bluetooth device states are:

- **Standby**: Waiting to participate in the piconet. This is a low power state.
- **Inquire**: Obtaining identification and clock offsets of the devices willing to be connected.
- **Page**: Making a connection after knowing the slave's identification and estimated clock offset.

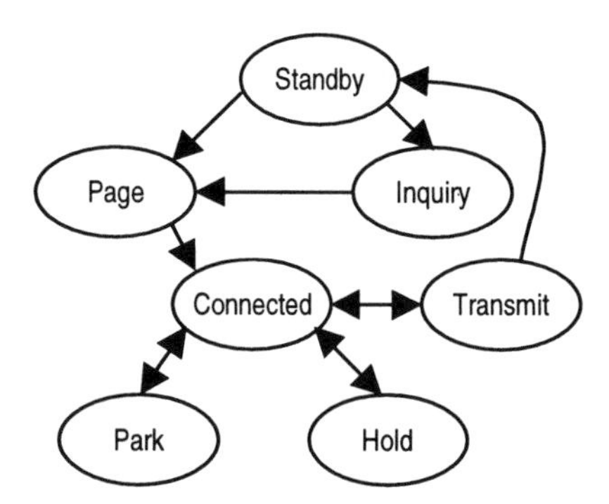

Figure 1-16. Bluetooth Device Operational States

- **Connected**: Active in a piconet as a master or a slave.
- **Hold**: The device has an active address and does not release the address. The device can resume sending after transitioning out of this state.
- **Park**: The device releases the address and remains synchronized with the piconet. The device listens to the traffic to re-synchronize and check for broadcast messages. The device has the lowest power consumption in this state.

1.11.3 Bluetooth SOC

Figure1-17 shows the block diagram of an example Bluetooth SOC design. It is based on an ARM7TDMI processor and AMBA. Refer to www.arm.com for more details on ARM7TDMI and AMBA.

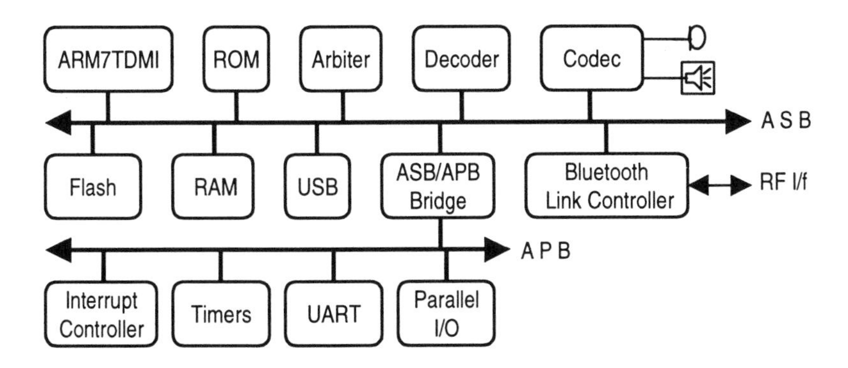

Figure 1-17. Block Diagram of Bluetooth SOC

1.11.3.1 Design Blocks

The Bluetooth SOC mainly consists of the following blocks:

- **ARM7TDMI processor**: Manages and schedules all activities. It receives the interrupt, stores data from input devices, processes data, and sets up operations for data transfer between memory and other devices accordingly. The processor runs on an RTOS, such as pSOS, VxWorks, or Windows CE. The processor also runs the complete Bluetooth protocol stack.
- **AMBA details**: The following buses are defined by AMBA, an on-chip bus architecture defined by ARM.

Advanced system bus (ASB): Interfaces high-bandwidth devices. The processor, memory, and high-speed peripherals (for example, USB, Codec, Bluetooth link controller) that are involved in a majority of data transfers are connected on the ASB.

Advanced peripheral bus (APB): Interfaces the devices that require lower bandwidth, such as timers, interrupt controller, universal asynchronous receiver and transimtter (UART), parallel I/O.

- **Arbiter**: Ensures that only one bus master initiates data transfers at a time.
- **Decoder**: Performs decoding of the transfer addresses and selects slaves appropriately.
- **Flash memory**: Holds the information of system configuration.
- **Random access memory (RAM)**: Holds the temporary data.
- **Read only memory (ROM)**: Holds the processor code for initialization and application program.
- **Bluetooth link controller**: Carries out the link controller functions and interfaces to the radio link. The functions include frequency hopping, channel access code generation, error correction, scrambling, authentication, encryption/decryption, and cyclic redundancy check as per the Bluetooth protocol standard.
- **Universal serial bus (USB)**: Connects peripherals to the processor. It can be used to attach a wide variety of devices like scanners, cameras, keyboards, and speakers.
- **Codec**: Provides an interface to the external world for audio data transfer. It connects to analog-to-digital converter (ADC) and digital-to-analog converter (DAC).
- **ASB to APB bridge**: This bridge is a slave on ASB and translates ASB transfers into a suitable format for the slave devices on the APB. It latches the address, data, and control, and performs address decoding to generate slave select signals for peripherals on APB. The interrupt controller, timers, parallel I/O, and UART are interfaced to APB.

  All the memories (ROM, RAM, and Flash), USB, Bluetooth link controller, Codec, and ASB/APB bridge are interfaced on an ASB.
- **UART**: Serial port that connects the peripheral capable of transferring the data in serial form.
- **Parallel I/O port**: 8-bit parallel I/O port that sets the outputs and reads the status of the input lines connected to it.

- **Interrupt controller**: Generates an interrupt to the processor upon a data transfer request from the peripherals. The interrupt output lines from all the peripherals are connected at the inputs of the interrupt controller.
- **Timers**: Two 16-bit timers used for timing applications.
- **Phased locked loop (PLL)**: Takes clock as input and generates appropriate clock signals required for the peripherals.

1.11.3.2 SOC Operation

The Bluetooth SOC explained above is programmable as a master or slave. The processor runs all the data handling, Bluetooth protocol stack, and control operations. Whenever the device is programmed as a master, the slave address and data transfer types are programmed by the processor. The voice signal generated by the microphone is converted to digital form by the ADC within the Codec block. Codec pre-processes the data and interrupts the processor. The processor, in its interrupt service routine, performs a data transfer from the Codec to RAM. Later the data from RAM is transferred to the Bluetooth link controller, which does the necessary data processing and sends the data to the RF link.

In receive mode, the Bluetooth link controller interrupts the processor whenever it receives the data from the RF link. The processor reads the data in the interrupt service routine and stores the data into RAM. Later, the processor reads the data and does pre-processing and passes the processed data to the Codec block. The DAC within the Codec block converts the data into analog and plays on the loud speaker. The Bluetooth SOC operates accordingly when programmed as a slave.

Summary

This chapter has provided an overview of the issues to consider and the various verification methodologies that can be used with SOC designs. We will now explore these topics in further detail in subsequent chapters.

References

1. Peters Kenneth H. Migrating to single-chip systems, Embedded Systems Programming, April 1999.

2. Geist Daniel, Biran Giora, ... Methodology for the verification of a "system on chip," Design Automation Conference 1999.

3. El-Ghoroury Hussein S. Next-generation IC designs for mobile handsets, IBM-MicroNews, second quarter 1999, Vol. 5, No. 2.

4. Goodnow Ken. Enhancing the design process using system-on-a-chip capability, IBM-MicroNews, first quarter 2000, Vol. 6, No. 1.

5. Siegmund Richard Jr. Shortened design cycles demand logic reuse, IBM-MicroNews, third quarter 1999, Vol. 5, No. 3.

6. Baker Mark, O'Brien-Strain Eamonn. Co-design made real: Generating and verifying complete system hardware and software implementations, Embedded Systems Conference, September 1999.

7. Leef Serge. Changing SOC content demands codesign and careful IP integration, Integrated System Design.

8. Reddy Anil. Integration of IP into SOC presents design challenge, Wireless Systems Design, December 1998.

9. Schirrmeister Frank, Martin Grant. Platform-based design helps meld EDA with convergence demands, Wireless Systems Design, May 2000.

10. Cooley David, Ostrowski Marc, ... System-on-a-chip designs calls for multiple technologies, Wireless Systems Design, July 1999.

11. Evans Adrian, Silburt Allan, ... Functional verification of large ASICs, Design Automation Conference 1998.

12. Tuck Barbara. Various techniques, languages being used to verify system-on-a-chip designs, Electronic Systems, January 1999.

13. Chapiro Daniel. Automating verification using testbench languages, Electronics Engineer, September 1999.

14. A variety of techniques/languages being used to verify SOC designs, Computer Design, January 1999.

15. Spec-based verification - A new methodology for functional verification of systems/ASIC, a whitepaper, www.verisity.com.

16. Quickbench verification suite, www.chronology.com.

17. Mitchell Donna. Test bench generation from timing diagrams, www.synapticad.com.

18. Verification with BestBench, A technical paper, www.diagonal.com.

19. Saunders Larry. Effective design verification, Integrated System Design, April 1997.

20. VERA - Testbench automation for functional verification, a technical paper, www.synopsys.com.

21. Tseng Ping-Sheng. Reconfigured engines rev simulation, EE Times, July 10, 2000.

22. Gallagher. Prototypes ensure pre-verification, EE Times, June 12, 2000.

23. Browne Jack. Tools take aim at system-level verification, Wireless Systems Design, June 2000.

24. Verification hardware takes center stage at DAC, EE Times, May 29, 2000.

25. Cheng Kwang-Ting, Dey Sujit, ... Test challenges for deep-submicron technologies, Design Automation Conference 2000.

26. Huott W V, Koprowski B J, ... Advanced microprocessor test strategy and methodology, Journal of Research and Development, Vol. 41, No. 4/5 - IBM S/390 G3 and G4, 1997.

27. Dey Sujit, Marinissen Eric Jan, Zorian Yervant. Testing system chips: Methodologies and experiences, Integrated System Design, 2000.

28. Comman Bejoy G. Design-for-test considerations challenge SOC developers, Wireless Systems Design, June 1999.

29. Diehl Stan. SOC designs demand embedded testing, Portable Design, March 2000.

30. Zorian Yervant, Marinissen Eric Jan, Dey Sujit. Testing embedded core-based system chips, Computer, June 1999.

31. Stannard David. Testing designs with embedded blocks, Electronics Engineer, May 2000.

32. Sunter Stephen. Testing mixed-signal ICs using digital BIST, Electronics Engineer, January 2000.

33. Embedded test solutions, adcBIST, memBIST-IC, logicBIST, www.logicvision.com.

34. Mark Olen, Rajski Janusz. Testing IP cores, Integrated System Design.

35. IEEE 1149.1 Boundary-scan in IBM ASICs, Application note, www.ibm.com.

36. Achieving DFT closure, A technical paper, www.synopsys.com.

37. Spaker Rebecca. Bluetooth Basics, Embedded Systems Programming, July 2000.

38. Kardach James. Bluetooth architecture overview, Intel Technology Journal Q2, 2000, www.intel.com.

39. Bluetooth protocol and product details, www.bluetooth.com.

40.Chang Henry, Cooke Larry, ... Surviving the SOC Revolution, A Guide to Platform-Based Design, Kluwer Academic Publishers, July 1999.

41. Keating Michael, Bricaud Pierre. Reuse methodology manual for system-on-a-chip designs, Kluwer Academic Publishers, 1999.

CHAPTER 2

System-Level Verification

In system-on-a-chip (SOC) design methodology, system design is performed after the system specification sign-off. In system design, the system behavior modeling is done and verified against the functional requirements. The behavior model is then mapped to an architecture comprised of intellectual property (IP) blocks. The system-level verification is performed to check the architecture against the intended functional and performance requirements. This chapter illustrates the following topics:

- System design
- System verification
- Creating system-level testbenches
- Applying and migrating testbench

Aspects of system-level verification are illustrated with the Bluetooth SOC design example introduced in Chapter 1.

2.1 System Design

In the system design process, the customer requirement specifications are captured and converted into system-level functional and performance requirements. The system design process does not have any notion of either hardware or software functionality or implementation to start with. The functional and performance analysis are performed to decompose the system-level description.

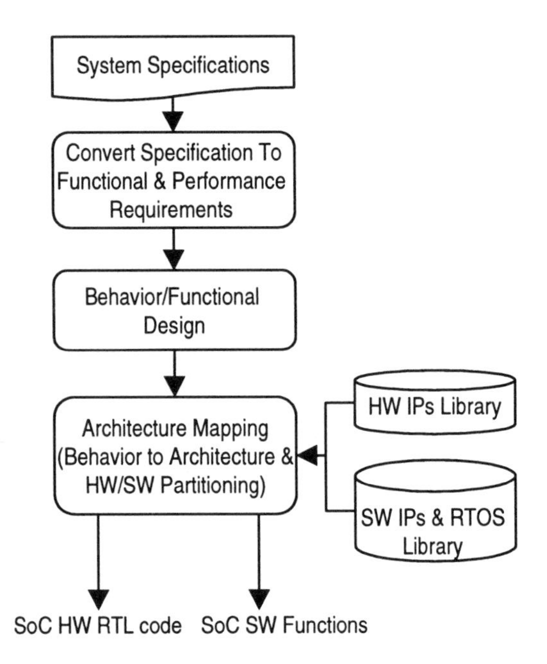

Figure 2-1. SOC System Design Flow

Figure 2-1 shows the system design methodology flow. In the functional design, a more detailed analysis of the processing requirements, data, and control flow for all the required functional modes of the system is performed. After the functional design, the behavior of the system is mapped to a candidate architecture of hardware and software elements available in the IP library. If the IPs are not available in the library, they need to be procured or developed to meet the design requirements. Abstraction models are used for IPs that are not available in the library during functional design.

The architecture selection is done based on trade-offs between the different alternatives. This is achieved by allocating the system-level processing requirements to hardware and/or software functions. The architecture is verified for intended functionality and performance. The hardware (SOC HW RTL code) and software (SOC SW functions) specification requirements are handed off to the respective design teams for detailed implementation and integration.

System design and verification can be performed using standard tools, such as Cadence® Virtual Component Co-Design (VCC), ConCentric, N2C, Cadence Signal Processing Worksystem, COSSAP, and others.

2.1.1 Functional/Behavioral Design

In the functional design, the algorithms are analyzed to assess the processing/computational, memory, and I/O requirements. The criteria to select an architecture are formalized, and the algorithm and process flows are translated to data and control flows that are independent of architecture.

2.1.2 Architecture Mapping

During architecture mapping and selection, different architectural design options and instantiations are evaluated. As shown in Figure 2-2, the following steps are involved:

- **Input**: The processing and functional requirements, IP library, criteria for selection, and data and control flow specifications of the system are considered as inputs for architecture selection and mapping.
- **Hardware IP library**: Includes behavioral models of the IPs, performance models, test plans, testbench sets, and documentation.
- **Software IP library**: Includes real-time operating system (RTOS) service routines, control software, device driver software, diagnostics software, application software or code fragments, such as encoders/decoders for moving picture experts group (MPEG) and joint photographic experts group (JPEG), encryption/decryption, compression/decompression, test data, and documentation
- **Partitioning**: The hardware/software partitioning and mapping for the candidate architecture is done.
- **Performance**: Performance analysis of the partitioning and mapping is carried out.
- **Optimization**: Mapping optimization is performed.

- **Analysis**: Analysis of architecture for size, power, testability, risk, reliability, and cost is performed
- **Selection**: The above steps are iterated until one or more acceptable architectures are selected. The hardware and software specifications of the architecture are handed off to the respective development teams.

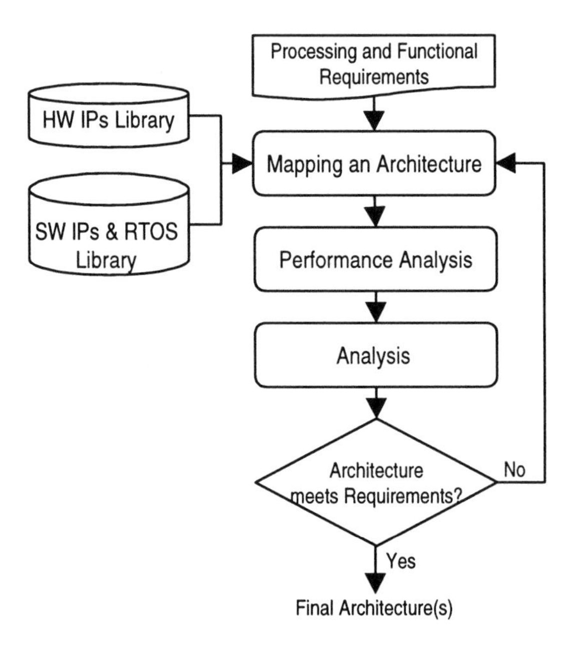

Figure 2-2. Architecture Mapping and Selection

2.2 System Verification

System-level verification is performed after functional design and architectural design and mapping. System-level testbenches are created, using system specifications as inputs, as shown in Figure 2-3.

During the functional design phase, the goal of the verification is to check whether the behavioral design meets the functional requirements. The verification and functional design are interactive processes that are performed until an acceptable functional design is selected.

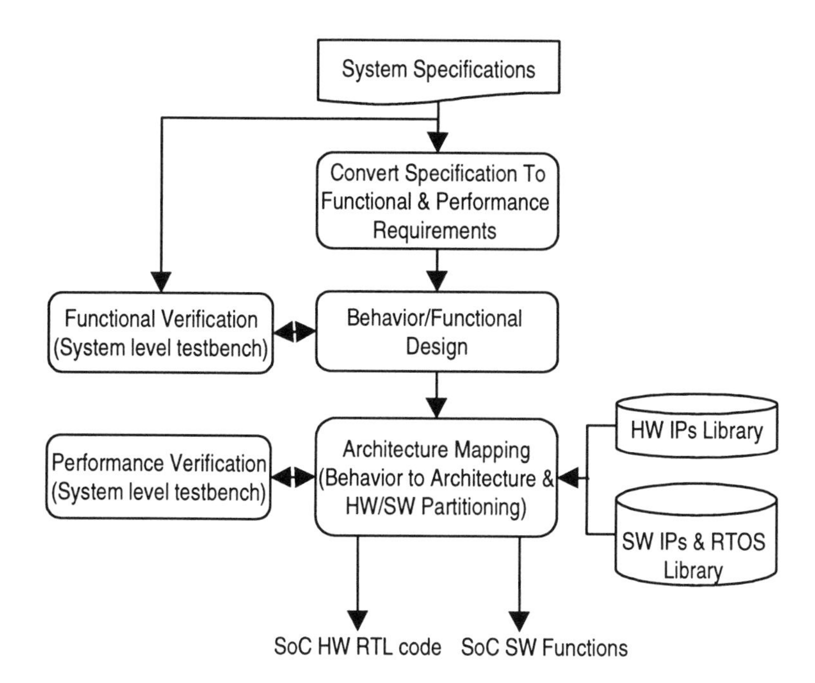

Figure 2-3. System-Level Design and Verification Flow

Performance verification is done during the architecture selection and mapping phase. In this phase, the goal is to check that the selected architectures meet performance requirements in addition to the functional requirements.

The following sections briefly illustrate functional and performance verification.

2.2.1 Functional Verification

Functional verification validates that the design meets the requirements. System-level testbenches are created based on the input specifications. The various aspects of the data and control flow are validated, which includes passing information between the external world, initiating or terminating I/O devices, and verifying software. The system-level testbenches created during the functional design can also be used for performance verification.

2.2.2 Performance Verification

Performance verification validates all the architectural entities and interfaces between them before detailed design implementation. All IPs, the optimized data, control flow details, and software description are used as inputs for the performance verification. The verification uses the simulators that are embedded in the system design environment. To verify the architecture, the testbench required at the system-level is created.

After performance verification, the output includes optimized library elements, and detailed functionality and performance specifications for hardware and software implementation.

2.2.3 System-Level Testbench

A system-level testbench is key to an overall top-down design and verification strategy. Testbenches should be created at the highest level of abstraction at which the design is modeled. The testbenches are used throughout the verification process to validate the functionality of the design as it proceeds through increasing levels of refinement and becomes more detailed.

2.2.4 Creating a System-Level Testbench

As shown in Figure 2-4, a system-level testbench is created by systematically extracting each functional requirement from the functional specification and defining a specific test to exercise that function.

Tests are performed on the functionality explicitly stated in the specification (for example, under conditions A, B, and C, the design will do W), as well as the implied functionality (for example, under conditions A, B, and C, the design will not do X, Y, or Z).

For each test, the success and failure conditions must be defined so that the functionality can be checked. For example:

- Data packet (xyz) will appear on port A
- No illegal bus cycles are generated during the execution of the test
- Memory location (xyz) contains value (AB) on completion of the test
- Variable (A) goes active during the execution of the test
- Variable (A) does not go active during the execution of the test

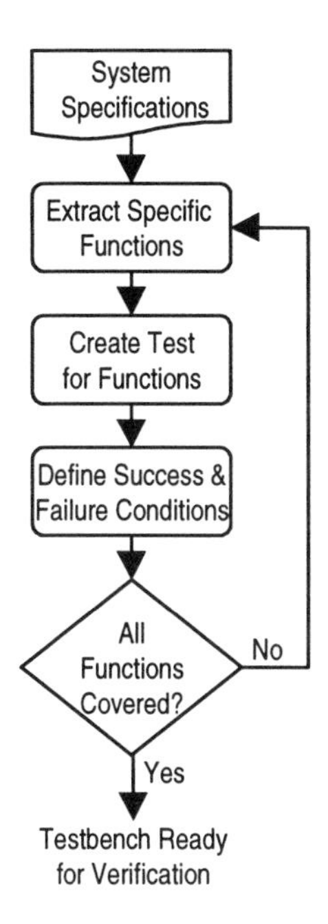

Figure 2-4. Creating a System-Level Testbench

To define and create a complete testbench, pay particular attention to the following:

- Corner cases
- Boundary conditions
- Design discontinuities
- Error conditions
- Exception handling

2.2.5 System Testbench Metrics

The metrics to be applied to a system testbench depend upon the level of abstraction at which the test is being evaluated. The first measure is whether or not all of the tests defined in the verification plan are included. This is a qualitative measure rather than a truly quantitative one. However, if attention was paid in the definition of the verification plan, this is a worthwhile metric.

At the abstract system level (C, C++, VCC, SystemC), the industry is interested in defining and measuring functional coverage that measures how well a testbench exercises a purely functional model of a design. While this area has attracted much interest, it has not yet resulted in definitive tools, although it does offer hope for the future. What can be done today is to identify the operations or transactions a design is capable of performing and to define the set of concurrent operations that will adequately test the design.

When possible, test all possible combinations of transactions with all possible data sets. For large designs, this results in excessively large testbenches. Apply knowledge of the design to limit the testbench size to a manageable level by excluding redundant tests without sacrificing the quality of test.

Query the simulation databases that log the results of multiple simulations to test if the transaction test requirements have been met.

At the hardware description language (HDL) level, a variety of tools measure code coverage. The types of metrics that can be quantified are statement, toggle, FSM arc, visited state, trigger, branch, expression, path, and signal coverages.

2.2.6 Applying the System-Level Testbench

The V-shaped model shown in Figure 2-5 describes a top-down design and bottom-up implementation approach to design.

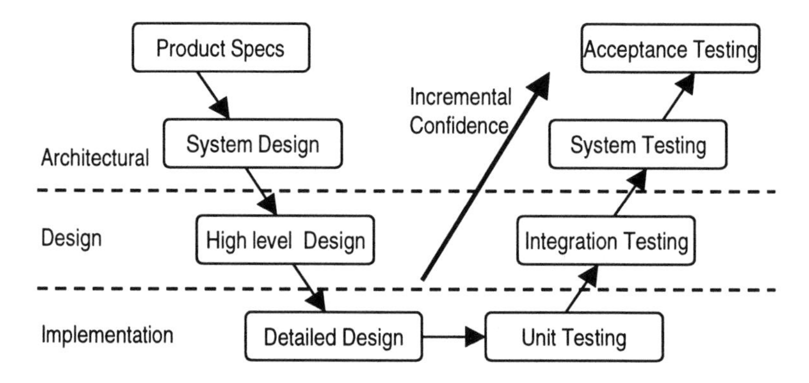

Figure 2-5. Top-Down Design and Bottom-Up Implementation Approach

At each stage in the design process, the system-level testbench is used to validate the integrated design. In practice, there are some challenges in applying a system testbench to a gate-level model of a complex SOC device. As the number of gates on an SOC device increases, the size of the system-level testbench required to test the SOC exhaustively grows exponentially. Software simulators continue to make impressive improvements in both overall capacity and run times, but they cannot keep up with this exponential growth.

To apply the complete system-level testbench to detailed implementations of each of the SOC elements, a number of approaches can be adopted. The following sections describe the various approaches using the SOC design shown in Figure 2-6 as an example.

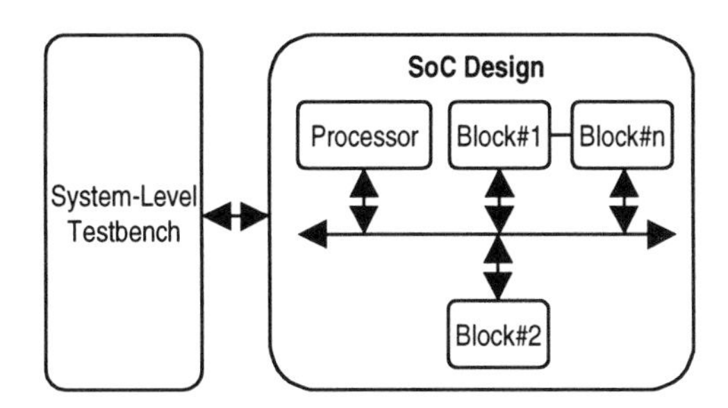

Figure 2-6. Verifying a System with Actual Blocks

2.2.6.1 Emulation

Emulators are specially designed hardware and software systems that contain some type of configurable logic, often field programmable gate arrays (FPGA). These systems are programmed to take on the behavior of the target design and can emulate its functionality, as shown in Figure 2-7.

Because these systems are hardware-based, they can provide simulation speeds of up to tens of megahertz, as opposed to tens of hertz to kilohertz speeds achievable with software simulators. This improvement in simulation speed enables emulation systems to execute the entire system test suite on the complete SOC design.

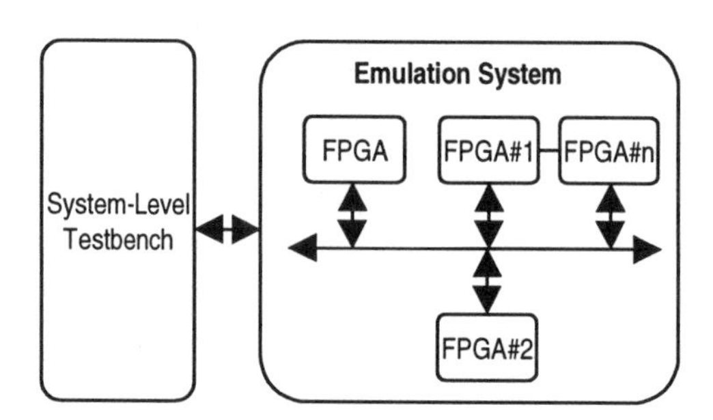

Figure 2-7. Verifying a System with Emulation

2.2.6.2 Hardware Acceleration

Hardware acceleration maps some or all of the components in a software simulation into a hardware platform specifically designed to speed up certain simulation operations. This speed-up in simulation performance enables the entire system-level testbench to be executed against the complete SOC design.

2.2.6.3 Hardware Modeling

A hardware modeler allows a bonded core for an IP to be substituted for the IP model in the simulation environment, as shown in Figure 2-8.

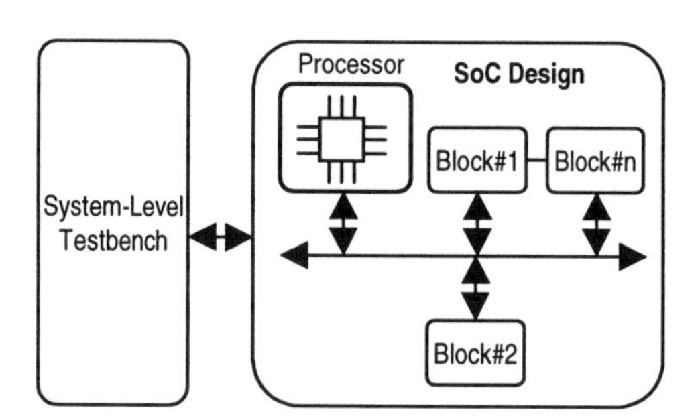

Figure 2-8. Verifying a System with a Processor Bonded Core

Because the core can run at real time, the overall simulation speed for the entire SOC is significantly improved. If the bonded cores loaded in the hardware modeler account for a significant amount of the overall design, this approach can support the simulation on the entire system-level testbench of the complete SOC design.

2.2.6.4 Mixed-Level Simulation

If the hardware speed-up techniques described above are not available to the design team, accelerate the software-based simulation by running mixed-level simulations. In a mixed-level simulation, different blocks of the design are executed at different levels of abstraction. Typically, the following two approaches can be used.

- If the design simulation speed is dominated by a single complex block, such as a processor core, substitute an abstract model for this block in the overall design, as shown in Figure 2-9.
- If the design has many complex blocks, run multiple simulations using abstract models for all but one of the blocks, as shown in Figure 2-10. For each simulation, a different block will have its detailed implementation view swapped into the simulation.

  With both of these approaches, the abstract models must be functionally equivalent to the implementation view. This is accomplished by either adopting the design partitioning approach described below or using abstract models certified by the block or virtual component (VC) supplier.

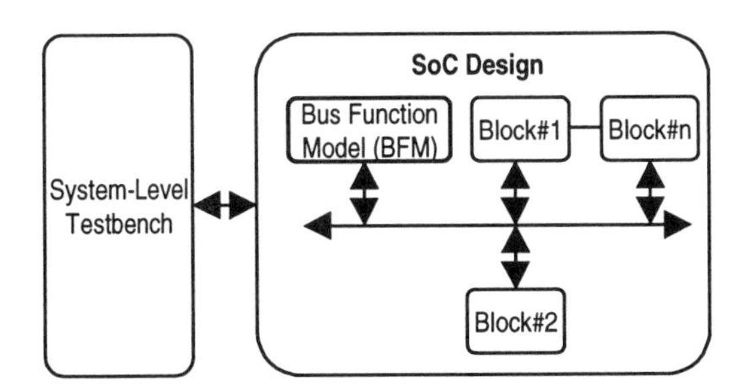

Figure 2-9. Verifying a System with a Processor BFM

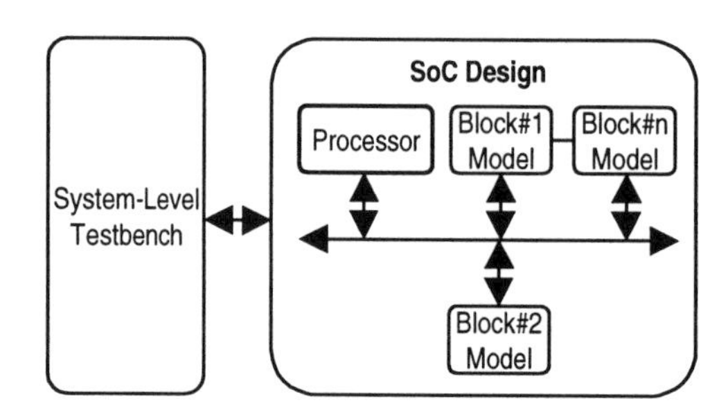

Figure 2-10. Verifying a System with Abstract Models for Blocks

2.2.6.5 Design Partitioning

After successfully executing the system-level testbench on a bit-accurate and cycle-accurate model, extract individual testbenches for each of the blocks or VCs within the design by placing probes on all of the I/Os to a block and capturing the input values as stimulus and the output values as expected results, as shown in Figure 2-11A. Apply the resulting file to the block in isolation, as shown in Figure 2-11B.

The verification at the block level is equivalent to the verification that would be performed on the block if the system testbench were applied on the entire SOC design. The detailed views of the blocks can then be verified in isolation using these block-level testbenches. This significantly reduces the simulation requirements to verify the entire SOC design.

2.2.7 System Testbench Migration

The top-down design and bottom-up implementation approach calls for the system-level testbench to be migrated across all views or abstraction levels of the design. To support this goal, methodologies must support the following two scenarios:

- Migration of a testbench between different abstraction levels of the design
- Migration of the system testbench between different environments at the same level of abstraction

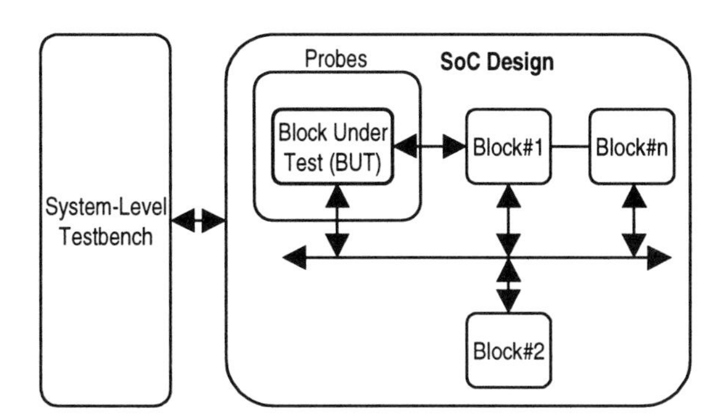

Figure A. Verifying Block in the System.

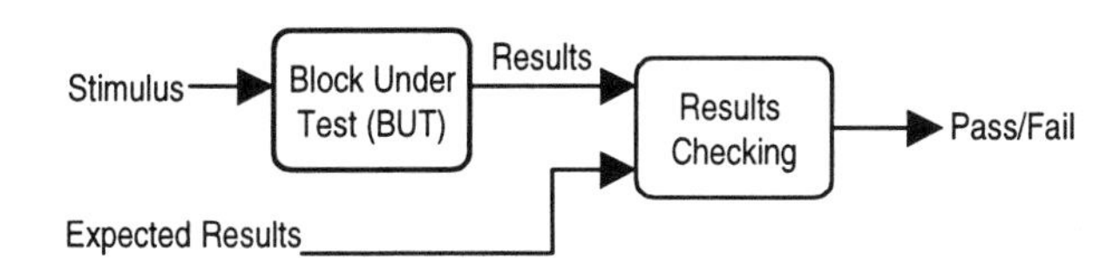

Figure B. Verifying Block in Isolation.

Figure 2-11. Verifying a Block in a System and in Isolation

2.2.7.1 Migrating a Testbench to Different Abstraction Levels

To migrate a testbench between different abstraction levels, identify the points of indifference between the models. Translate the testbench from the higher abstraction level to the lower one. For example, a functional testbench with data tokens or packets must be translated into a bus-accurate and cycle-accurate testbench. Then apply the testbenches to the associated model and compare the results at the points of indifference to ensure functional equivalence.

To account for the increased functional detail, enhance the lower-level testbench. If a communications protocol was introduced in the transition from one level to the next, additional tests are required to fully test the implementation of that protocol. If a test interface and controller have been introduced, they must be fully tested.

2.2.7.2 Migrating a System Testbench to Different Environments

Migrating the system testbench between different environments at the same abstraction level is a two-step process. First, check the original testbench format and contents to ensure that they contain all of the information required for the new environment. Second, translate the original testbench into a format compatible with the new environment.

2.3 Bluetooth SOC

System design for the example Bluetooth SOC design is performed using Cadence's VCC environment.

Figure 2-12A shows the behavioral/functional model, which consists of a link manager and Bluetooth link controller models. The link manager creates protocol messages for link set-up, security, and control. The Bluetooth link controller carries out the baseband protocol and other low-level link routines. The link manager model is driven by a system-level testbench. Figure 2-12 shows two Bluetooth devices: #1 is transmitting, and #2 is receiving. The output of #1 is connected as input to #2 for illustration purposes.

Testbench #1 drives the link manger in device #1 with data that is to be transmitted by the associated link controller. Testbench #2 drives the link manager in device #2 to receive the data that is available from the associated link controller.

Figure 2-12B shows the architecture mapping of Bluetooth device #1. The behavior model is mapped to an architecture consisting of a CPU (ARM7TDMI), associated memory, a Bluetooth link controller, and an RTOS. The link manager and part of the Bluetooth link controller are mapped to the software running on the ARM7TDMI.

The results of the performance analysis are shown in Figure 2-13. Figure 2-13A shows the CPU utilization, which is 0.00272 for a 25MHz clock speed. Figure 2-13B shows the bus mean utilization, which is 0.0008.

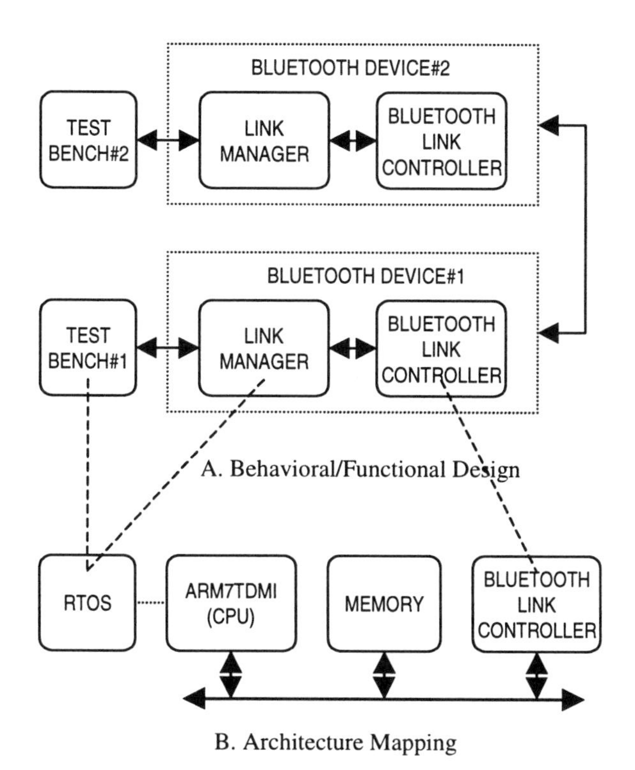

Figure 2-12. Bluetooth Functional Design and Architecture Mapping

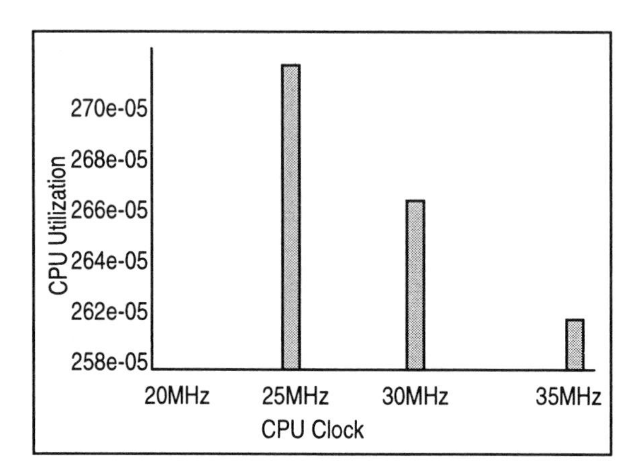

A. CPU Utilization Graph

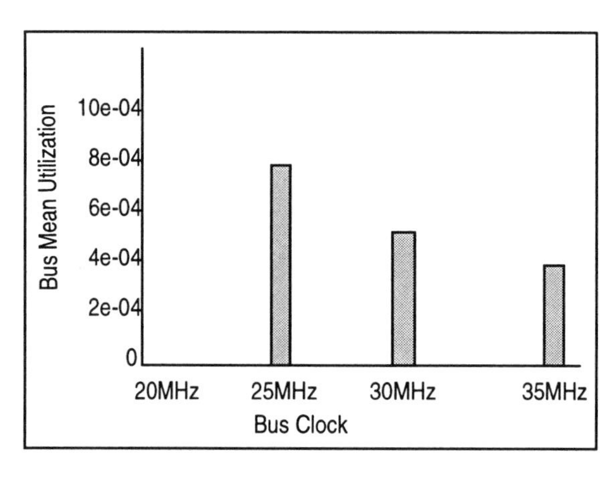

B. Bus Mean Utilization Graph

Figure 2-13. CPU and Bus Mean Utilization Graphs

The following examples show the pseudocode for the link manager and link controller models, and the system-level testbench.

Example 2-1. Pseudocode for Link Manager Model

```
#include "black.h"

CPP_MODEL_IMPLEMENTATION::CPP_MODEL_IMPLEMENTATION(cons
t ModuleProto &proto, InstanceInit &inst)
: CPP_MODEL_INTERFACE(proto, inst)
{
}

void CPP_MODEL_IMPLEMENTATION::Init()
{
 /* Contains the initialization code for Link Manager
functional block */
}

// Should you choose to use Await(), uncomment the fol-
// lowing member function definition and uncomment the
// declaration for it within the .h file. It will be
// called after the beginning of the simulation
// as soon as the block gets scheduled.

// N.B., this function should never return.

// N.B., be sure to set the functional package parameter
// for this block named "UsesAwait" to "@VCC_Types.Yes
// NOType::Yes".

//void CPP_MODEL_IMPLEMENTATION::Begin()

//{
//}

// Should you choose to use Await (see above), this mem-
// ber function will never be called unless your Begin()
// member function calls it explicitly.
```

```
// We suggest that you comment it out, but this is not
// required.

// Should you choose not to use Await(), you must pro-
// vide a definition for this member function. It will
// be called each time the block reacts.
//

void CPP_MODEL_IMPLEMENTATION::Run()
{
/* Contains the behavioral code for the Link Manager */
/* Translate the HCI commands to LMP messages and pass
on to Link Controller */
}
```

Example 2-2. Pseudo Code for Link Controller Model

```
#include "black.h"

CPP_MODEL_IMPLEMENTATION::CPP_MODEL_IMPLEMENTATION(cons
t ModuleProto &proto, InstanceInit &inst)
: CPP_MODEL_INTERFACE(proto, inst)
{

}

void CPP_MODEL_IMPLEMENTATION::Init()
{
 /* Contains the initialization code for Link Controller
functional block */
}


// Should you choose to use Await(), uncomment the fol-
// lowing member function definition and uncomment the
// declaration for it within the .h file. It will be
// called after the beginning of the simulation as soon
```

```
// as the block gets scheduled.

// N.B., this function should never return.

// N.B., be sure to set the functional package parameter
// for this block named "UsesAwait" to "@VCC_Types.Yes-
// NOType::Yes".

//
//void CPP_MODEL_IMPLEMENTATION::Begin()
//{
//}

// Should you choose to use Await (see above), this mem-
// ber function will never be called unless your Begin()
// member function calls it explicitly.

// We suggest that you comment it out, but this is not
// required.

// Should you choose not to use Await(), you must pro-
// vide a definition for this member function. It will
// be called each time the block reacts.
//


void CPP_MODEL_IMPLEMENTATION::Run()
{

/* Contains the behavioral code for the Link Controller
including the state machine representing various connec-
tion states */
/* Packetize the messages and pass on to the RF */

}
```

Example 2-3. Pseudocode for System-Level Testbench

```
#include "black.h"

CPP_MODEL_IMPLEMENTATION::CPP_MODEL_IMPLEMENTATION(cons
t ModuleProto &proto, InstanceInit &inst)
: CPP_MODEL_INTERFACE(proto, inst)
{
}

void CPP_MODEL_IMPLEMENTATION::Init()
{
/* Contains the initialization code for the test bench
*/
}

// Should you choose to use Await(), uncomment the fol-
// lowing member function definition and uncomment the
// declaration for it within the .h file. It will be
// called after the beginning of the simulation as soon
// as the block gets scheduled.

// N.B., this function should never return.

// N.B., be sure to set the functional package parameter
// for this block named "UsesAwait" to "@VCC_Types.Yes-
// NOType::Yes".

//
//void CPP_MODEL_IMPLEMENTATION::Begin()
//{

//}

// Should you choose to use Await (see above), this mem-
// ber function will never be called unless your Begin()
// member function calls it explicitly.
```

```
// We suggest that you comment it out, but this is not
// required.

// Should you choose not to use Await(), you must pro-
// vide a definition for this member function. It will
// be called each time the block reacts.
//

void CPP_MODEL_IMPLEMENTATION::Run()
{
/* Contains the code for Bluetooth protocol testbench
(HCI level) */
/
********************************************************
Initiate the connection
Accept the connection confirmation
Initiate the ACL connection
Accept the ACL connection confirmation
DATA transfer ..........................................
Terminate ACL connection
End of testbench
*******************************************************/
}
```

Summary

System behavior modeling and verification are the first steps in implementing an SOC design. Developing a system-level testbench is essential for validating the design's functionality at each stage in the design process. Different techniques can be used when applying a testbench to a complex SOC device.

References

1. Corman, Tedd. Verify your system-level designs with a virtual prototype, Electronic Design, October 18, 1999.

2. Gupta Rajesh K, Zorian Yervant. Introducing core-based system design, IEEE Design & Test of Computers, October-December 1997.

3. Leef, Serge. Meeting the challenges of co-design, Electronics Engineer, March 2000.

4. Rincon Ann Marie, Lee William R, Slattery Michael. The changing landscape of system-on-a-chip design, IBM - MicroNews, third quarter 1999, Vol. 5, No. 3.

5. Lynch John, Schiefer Harold. Concurrent engineering delivers at the chip and system level, Integrated System Design, December 1997.

6. Giest Daniel, Biran Giora. Tackling the system verification of a network router, Integrated System Design, June 1999.

7. Bassak Gil. Focus report: Electronic system-level design tools, Integrated System Design, April 1998.

8. Cassagnol Bob, Weber Sandra, ... Codesigning a complete system on a chip with behavioral models, Integrated System Design, November 1998.

9. Ajluni Cheryl. True co-design is still over the horizon, Embedded System Development, March 2000.

10. Keating Michael, Bricaud Pierre. Reuse methodology manual for system-on-a-chip designs, Kluwer Academic Publishers, 1999.

11. Chang Henry, Cooke Larry, ... Surviving the SOC Revolution, A Guide to Platform-Based Design, Kluwer Academic Publishers, July 1999.

12. Cadence VCC Users and Reference manuals, www.cadence.com.

CHAPTER 3

Block-Level Verification

As system-on-a-chip (SOC) gets more complex, integrating intellectual property (IP) blocks into the design becomes a critical task in the SOC design methodology flow. The block must be verified before it is integrated with the system to make sure that it fits into the intended design.

This chapter addresses the following topics:

- Types of IP blocks
- Block-level verification
- Lint checking
- Formal model checking
- Functional verification
- Protocol checking
- Directed random testing
- Code coverage analysis

Block-level verification is illustrated with the blocks used in the Bluetooth SOC design example.

3.1 IP Blocks

IPs are either brought in from within the company or licensed from third-party sources. IPs are generally designed for use in a range of applications. The environment around the IP depends on the application in which the IP is being used and the other blocks that interface to the IP.

IP blocks used in SOC are available in the following forms:

- **Digital and analog/mixed signal (AMS) hardware blocks**

  Hardware blocks can be hard, firm, or soft. Hard cores are representations with placement and routing already done and targeted to a particular technology and library. The IP provider verifies them for functionality and performance. They are mainly implemented in standard-cell and full-custom.

  Firm cores are semi-hard cores and are user-configurable if required. They are provided in a package that contains core specifications, synthesizable register-transfer level (RTL) code, gate-level netlist, pre-placement information, estimated performance and gate count, and testbenches to verify the core.

  Soft cores are available as synthesizable RTL code or gate-level netlist. They are supplied along with a testbench. They can be targeted for any technology and library. The timing information for soft cores can be determined only after the detailed implementation of the core is complete.

  IP providers can supply these cores along with higher abstraction level models. This helps in verifying the core with the testbench supplied.

- **Software blocks**

  Software IPs are not as available in the industry as hardware IPs, because many of the software IPs are developed by design houses as needed. Also, matured software IP standards do not yet exist.

  The software blocks are mainly generic device drivers, real-time operating system (RTOS) kernels, and application algorithms. The generic device drivers are developed for the standard peripherals, such as universal asynchronous receiver and transmitters (UART), universal serial buses (USB), fire-wires, general purpose parallel input/output ports (GPIO), liquid crystal displays (LCD), keyboards, mouses, Codecs, and others. The device driver routines are recompiled for the processor used in the design and verified for correct functions.

  The application algorithms include fast fourier transform (FFT), filters, compression/decompression, encryption/decryption, joint photographic experts group (JPEG) and moving picture experts group (MPEG) algorithms, error detecting, correcting codes, and others. The algorithms are recompiled for the processor used in the design and verified for correct functions.

3.2 Block Level Verification

Figure 3-1 shows the methodology flow for hardware block-level verification. The methodology assumes that the block RTL code is the input.

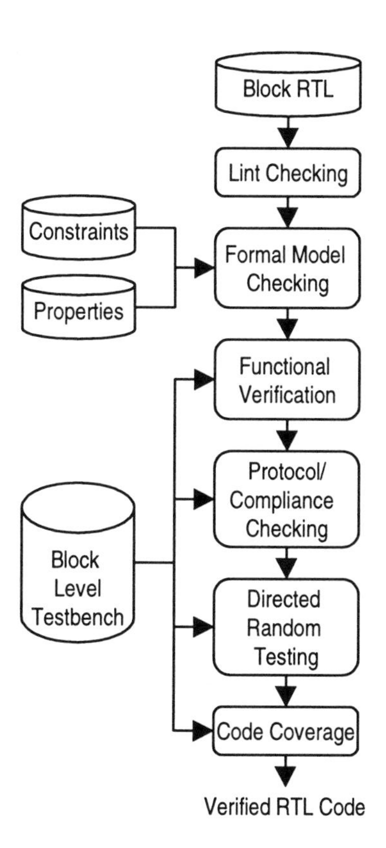

Figure 3-1. Block-Level Verification Flow

The RTL code goes through lint checking for syntax and synthesizability check. Formal model checking is performed to verify behavioral properties of the design. Model checking tools use the design's constraints and properties as input.

The RTL functional simulation uses the block-level testbench. Event-based or cycle-based simulation can be run on the block, depending on the simulation requirements.

Protocol/compliance testing verifies the on-chip and interface bus protocols. Directed random testing checks the corner cases in the control logic. These use probabilistic distributing functions, Poisson and uniform, which enable simulation using real-life statistical models. Code coverage identifies untested areas of the design.

Some of the IP providers verify the designs using rapid prototyping or an emulation system.

3.3 Block Details of the Bluetooth SOC

Figure 3-2 shows a simple block diagram of the example Bluetooth SOC design (details of the Bluetooth SOC example are explained in Chapter 1). The design blocks used to illustrate block-level verification are highlighted.

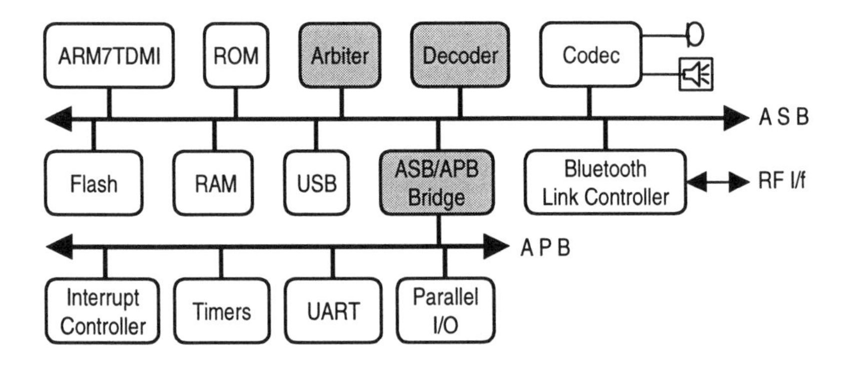

Figure 3-2. Block Diagram of Bluetooth SOC

Lint checking, testbench creation, and code coverage are illustrated with the arbiter block. Protocol/compliance checking is illustrated with the decoder and advanced system bus (ASB) master blocks. Directed random testing is described with the decoder block. Formal model checking is described with the arbiter, ASB/APB bridge, and decoder blocks.

3.3.1 Arbiter

The arbiter accepts the request from the bus masters and ensures that only one bus master is allowed to initiate data transfers at a time. The arbiter used in the design handles four bus masters, however, the design example explained here shows the design for two masters on the ASB for simplicity.

The arbiter is incorporated with round-robin and fixed priority-based schemes for bus access to the masters. A particular scheme can be programmed as required. In a round-robin scheme, the bus access is given to the requesting master based on least priority to the most recently serviced bus master. In a fixed-priority scheme, every master is programmed with its own priority.

Figure 3-3 shows a simple block diagram of the arbiter used in the Bluetooth SOC design. The requesting master generates a bus request signal on the areq line connected to it. Upon detecting the bus request, the arbiter grants the bus to the requesting master as per the priority scheme programmed. The arbiter indicates the grant to the master by asserting the 'agnt' signal connected to the requesting bus master. The master acquires the bus and performs the data transfer.

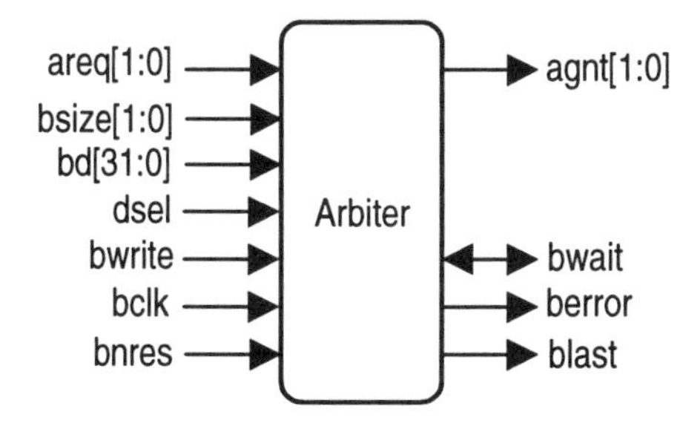

Figure 3-3. Block Diagram of the Arbiter

The following is a list of the arbiter block's I/O signals:

- areq—Bus master request line
- agnt—Bus grant line to the requested bus master
- Input control lines:

  bwrite—Read(0)/write(1) line generated by the current bus master

  bsize—Number of bytes indicated by the current bus master

  bclk—Bus clock signal line

  bnres—Reset signal line

  dsel—Slave select line

  bd—Data bus lines (bidirectional bus)
- I/O control signal lines:

  bwait—Signal generated by the selected device requesting wait state insertion

  blast—Last data transfer indication signal

  berror—Bus error indication signal

Figure 3-4 shows the state diagram of the arbiter block.

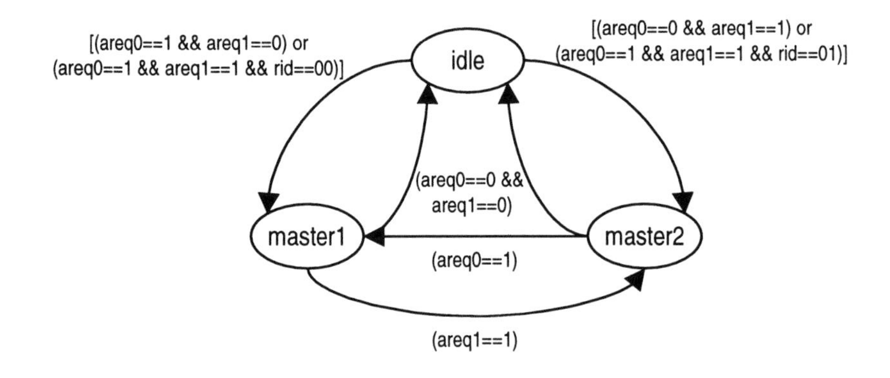

Figure 3-4. Arbiter Block State Diagram

The RTL code for the arbiter block is shown in Example 3-1.

Example 3-1. RTL Code for the Arbiter Block

```
module arbiter
(
 bnres, bclk, areq0, areq1, agnt0, agnt1, bwrite,
 bsize, bwait, blast, berror, bd, dsel
);

input          bnres; // Bus reset signal input
input          bclk;   // Bus clock signal
input          areq0;  // Master 0 bus request line
input          areq1;  // Master 1 bus request line
input          dsel;   // Device select line
input          bwrite; // Read/write line
input [1:0]    bsize;  // Number of bytes indication

inout          bwait;  // Wait state insertion request
inout          blast;  // Last data transfer indication
inout          berror; // Bus error indication
inout   [31:0] bd;      // Data bus

output          agnt0;  // Master 0 grant line
output         agnt1; // Master 1 grant line
// reg declaration
reg            agnt0;
reg            agnt1;
reg    [31:0] bd_r;
reg    [31:0] areg; // Arbiter Register
reg    [1:0]  rid;
reg            bwait_r;
reg            blast_r;
reg            berror_r;
reg            dset;
reg    [1:0]  agnt;
reg     [1:0] current_state, next_state; /*synopsys enum
code*/

// Internal wire signal declaration
wire   [31:0] bd = (!bwrite && dsel) ? bd_r : 32'hZ;
```

```
wire          bwait = (bclk)? 3'bzzz : bwait_r;
wire          blast = (bclk)? 3'bzzz : blast_r;
wire          berror = (bclk)? 3'bzzz : berro_r;

// State encoding
 parameter [1:0] /*synopsys enum code*/
 idle   = 2'b00,
 master1 = 2'b01,
 master2 = 2'b10;

/* Next State Block for encoding round-robin Priority */
//synopsys state_vector current_state

always @(current_state or areq1 or areq0 or rid)
begin
 next_state = idle;
  case (current_state)
   idle:
        if (areq1 == 1 && areq0 == 0)
         next_state = master2;
        else if (areq0 == 1 && areq1 == 0)
         next_state = master1;
        else if (areq0 ==1 && areq1 ==1 && rid ==2'b00)
         next_state = master1;
        else if (areq0 ==1 && areq1 ==1 && rid ==2'b01)
         next_state = master2;
        else next_state = idle;
   master1: if (areq1 == 1)
             next_state = master2;
            else  if (areq1 == 0 && areq0 == 0)
             next_state = idle;
            else
              next_state = master1;
   master2: if (areq0 == 1)
             next_state = master1;
       else if (areq1 == 0 && areq0 == 0)
        next_state = idle;
       else
        next_state = master2;
  endcase
end //Next state Block
```

```
/*** Output Block *****/
always @(current_state or areq0 or areq1)
begin
 case (current_state)
   master1 : begin
              agnt0 = 1'b1;
              dset  = 1'b1;
               casex ({dset,rid,areq0})
                4'b0xx_1 : begin
                            agnt0 = 1'b1;
                            dset  = 1'b1;
                           end
                4'b101_1 : begin
                            agnt1 = 1'b0;
                            dset  = 1'b0;
                            rid   = 2'b00;
                           end
                default: ;
               endcase
             end
   master2 : begin
              agnt1   = 1'b1;
              dset    = 1'b1;
               casex ({dset,rid,areq1})
                4'b0xx_1 : begin
                            agnt1 = 1'b1;
                            dset  = 1'b1;
                           end
                4'b100_1 : begin
                            agnt0 = 1'b0;
                            dset = 1'b0;
                            rid  = 2'b01;
                           end
               endcase
            end
 endcase
end //  Output Block

always @(negedge bclk or negedge bnres)
begin
 if (bnres == 1'b0)
```

```
  begin
   current_state <= idle;
   dset <= 1'b0;
   rid  <= 2'b00;
  end
 else
  current_state <= next_state;
end

//Bus Signal Assignments
always @(negedge bclk or negedge bnres)
 begin
  if (bnres == 1'b0)
    {bwait_r,blast_r,berro_r} <= 3'bzzz;
  else
    if (dsel )
      {bwait_r,blast_r,berro_r} <= 3'b000;
    else
      {bwait_r,blast_r,berro_r} <= 3'bzzz;
 end

always @(negedge bclk or negedge bnres)
 begin
  if (bnres == 1'b0)
   begin
    areg <= 32'h0000_00E4;
            // Fixed: 11 <- 10 <- 01 <- 00
    bd_r <= 32'hzzzz_zzzz;
    rid  <= 2'b00;
    dset <= 1'b0;
    agnt <= 2'b00;
   end
  else
   begin
    case ({dsel,bwrite})
     2'b11 :
            areg    <= bd;
     2'b10 :
            bd_r    <= areg;
     default:
            bd_r    <= 32'hzzzz_zzzz;
```

```
    endcase
   end
  end // always block
endmodule
```

3.3.2 Arbiter Testbench

The arbiter testbench tests the functionality of the arbiter. The testbench generates the input stimulus to the arbiter, collects the output responses, and compares them with the expected response. The stimulus is applied to the arbiter by instantiating it in the testbench module.

The following aspects of the arbiter are tested using the testbench:

- Reset functionality
- Whether the bus is granted to master1 and master2 upon request
- Effect of ‘blok’ signal
- Arbitration when both masters request the bus

3.3.2.1 Verilog Testbench

Example 3-2 shows the testbench code for the arbiter in Verilog HDL.

Example 3-2. Testbench for the Arbiter Block

```
module arbiter_test;
reg bclk, areq0,areq1;
reg bwrite;
reg [1:0] bsize;
wire [31:0] bd;

arbiter ARB(
 bnres, bclk, areq0, areq1, agnt0, agnt1, bwrite,
 bsize, bwait, blast, berror, bd, dsel
);

initial
begin
  bclk = 0;
  bnres = 0; #50;
```

```
  $display ("\n------------------------------------");
  $display ("\nArbiter simulation");
  $display ( "***************\n");
  #200;
  areq0 = 0; areq1 = 1;
  #20;
  areq0 = 1; areq1 = 0;
  #60;
  areq0 = 0; areq1 = 1;
  #90;
  areq0 = 1; areq1 = 0;
  #110;
  areq0 = 0; areq1 = 0;
  #510;
  areq0 = 1; areq1 = 1;
  #90;
  areq0 = 1; areq1 = 0;
  #100 $stop;
end
always
  bclk = #20 ~bclk;
endmodule
```

3.3.2.2 PLI Testbench

The programming language interface (PLI) extends the Verilog HDL by allowing user-defined utilities to access the design. The utilities can be monitoring tasks, stimulus tasks, debug tasks, translator tasks, delay calculators, and so on. The user-defined tasks are linked to a C function, which can then be integrated in Verilog using PLI. An application can read and write values on signals declared in Verilog. VHDL also supports a similar interface that includes C routines.

For example, a task can be defined in C and later called in Verilog code. Also, the tasks defined in C can be called from the command line during simulation. PLI applications can be dynamically or statically linked to the simulator. When linked dynamically, the simulator is run in normal way. When linked statically, the simulator is run with the customized executable file that is created. Example 3-3 gives a C function written for u-law to linear format conversion and called in Verilog testbench.

Example 3-3. PLI Testbench

```
double u_to_lin(pcm)
unsigned char   pcm;
{
    long    q, s, sgn, mag;
     /* process u-law pcm */
    pcm ^= 0xff;
    q = 0x0f & pcm;
    s = (0x70 & pcm) >> 4;
    sgn = 0x80 & pcm;
    mag = (((2 * q) + 33) << s) - 33;
    mag = (sgn ? mag | (1 << 13) : mag) & 16383;
    return((double) mag);
}
```

The task defined in the above C function is called in the Verilog testbench as follows:

```
module u_to_lin_test;
integer fg;
initial
begin
  fg = $fopen ("out.txt");
  $u_to_lin ("data.txt");
end
endmodule
```

3.3.2.3 Waveform-based Testbench

Signal waveforms are edited according to the design requirement. A tool that converts the waveforms into a stimulus generation embedding the timing information is used to create the testbench. The testbench is then used to verify the DUT.

Figure 3-5 shows an example waveform for the arbiter block. The input signals simulated are clock (bclk), bus request signals (areq0 and areq1), and reset signal (bnres).

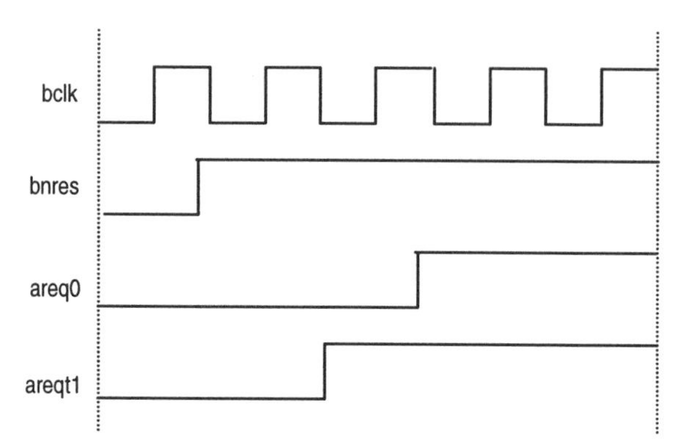

Figure 3-5. Waveform-based Testbench

When the input waveforms are fed to a hypothetical waveform-based testbench generation tool, it results in the testbench shown in Example 3-4.

Example 3-4. Code Generated by a Waveform-based Testbench Generation Tool

```
'timescale 1ns/1ns
module testbench;
reg bclk, bnres, areq0, areq1, dsel bwrite ;
reg [1:0] bsize;
wire bwait, blast, berror;
wire [31:0] bd;
wire agnt0, agnt1;

defparam DUT.idle = 2'b00;
defparam DUT.master1 = 2'b01;
defparam DUT.master2 = 2'b10;

arbiter DUT (
.areq0(areq0),          .areq1(areq1),          .dsel(dsel),
.bwrite(bwrite),        .bsize(bsize),        .bwait(bwait),
.blast(blast), .berror(berror), .bd(bd), .bnres(bnres),
.bclk(bclk), .agnt0(agnt0), .agnt1(agnt1)
            );
integer TX_FILE, TX_ERROR;
initial
```

```
begin
TX_ERROR=0;
TX_FILE=$fopen("results.txt");
areq0 = 1'b0;
areq1 = 1'b0;
asel = 1'b0;
bwrite = 1'b0;
bsize = 2'b00; //0
anres = 1'b0;
bclk = 1'b1;

#10 bclk = 1'b0; #10
#40 blck = 1'b1;
#40 bnres = 1'b1;
#10 blck = 1'b0; #10
#40 blck = 1'b1;
#40 areq1 = 1'b1;
#10 bclk = 1'b0; #10
#40 bclk = 1'b1;
#40 areq0 = 1'b1;
areq1 = 1'b0;
#10 bclk = 1'b0; #10
#40 bclk = 1'b1;
#40 areq0 = 1'b0;
areq1 = 1'b1;
#10 bclk = 1'b0; #10
#40 bclk = 1'b1;
#40 areq0 = 1'b1;
areq1 = 1'b0;
#10 bclk = 1'b0; #10
#40 bclk = 1'b1; #40
#10 bclk = 1'b0; #10
#40 bclk = 1'b1;
#40 areq0 = 1'b1;
areq1 = 1'b1; #10
if (TX_ERROR == 0) begin
$display("No errors or warnings");
$fdisplay(TX_FILE,"No errors or warnings");
end else begin
$display("%0d errors found in simulation",TX_ERROR);
```

```
$fdisplay(TX_FILE,"%0d   errors   found   in   simula-
tion",TX_ERROR);
end

$fclose(TX_FILE);
$stop;
end

task check_bwait;
input next_bwait;
#0 begin
if (next_bwait !== bwait) begin
$display("Error at time=%0dns bwait=%1b, expected=%1b",
$time, bwait, next_bwait);
$fdisplay(TX_FILE,"Error   at   time=%0dns   bwait=%1b,
expected=%1b",
$time, bwait, next_bwait);
TX_ERROR = TX_ERROR + 1;
end
end
endtask

task chexk_blast;
input next_blast;
#0 begin
if (next_blast !== blast) begin
$display("Error at time=%0dns blast=%1b, expected=%1b",
$time, blast, next_blast);
$fdisplay(TX_FILE,"Error   at   time=%0dns   blast=%1b,
expected=%1b",
$time, blast, next_blast);
TX_ERROR = TX_ERROR + 1;
end
end
endtask

task check_berror;
input next_berror;

#0 begin
if (next_berror !== berror) begin
```

```
$display("Error at time=%0dns berror=%1b,
expected=%1b",
$time, berror, next_berror);
$fdisplay(TX_FILE,"Error at time=%0dns berror=%1b,
expected=%1b",
$time, berror, next_berror);
TX_ERROR = TX_ERROR + 1;
end
end
endtask

task check_bd;
input [31:0] next_bd;
#0 begin
if (next_bd !== bd) begin
$display("Error at time=%0dns bd=%32b, expected=%32b",
$time, bd, next_bd);
$fdisplay(TX_FILE,"Error at time=%0dns bd=%32b,
expected=%32b",
$time, bd, next_bd);
TX_ERROR = TX_ERROR + 1;
end
end
endtask

task check_agnt0;
input next_agnt0;
#0 begin
if (next_agnt0 !== agnt0) begin
$display("Error at time=%0dns agnt0=%1b, expected=%1b",
$time, agnt0, next_agnt0);
$fdisplay(TX_FILE,"Error at time=%0dns agnt0=%1b,
expected=%1b",
$time, agnt0, next_agnt0);
TX_ERROR = TX_ERROR + 1;
end
end
endtask

task check_agnt1;
input next_agnt1;
```

```
#0 begin
if (next_agnt1 !== agnt1) begin
$display("Error at time=%0dns agnt1=%1b, expected=%1b",
$time, agnt1, next_agnt1);
$fdisplay(TX_FILE,"Error   at   time=%0dns   agnt1=%1b,
expected=%1b",
$time, agnt1, next_agnt1);
TX_ERROR = TX_ERROR + 1;
end
end
endtask
endmodule
```

3.3.2.4 Testbenches with Timing

In testbenches with timing information, delays can be assigned to paths across the module and timing checks can be performed to ensure that the timing constraints for the design are met. Example 3-5 shows the path delays defined and the timing checks that are performed in a testbench.

Example 3-5. Testbench with Timing

```
specify
   //three specparam declarations
   specparam tRise_clk_agnt0 = 150:200:250,
             tFall_clk_agnt0=200:250:300;
   specparam tSetup=60:70:75, tHold=45:50:55;
   specparam tWpos=180:600:1050, tWneg=150:500:880;
   //path assignment
   (bclk=>agnt0) = (tRise_clk_agnt0, tFall_clk_agnt0);

   //System timing checks
   //Setup time for Areq0 to posedge of clk
   $setup(areq0, posedge bclk, tSetup);
   // hold time for posedge of clk to areq0
   $hold(posedge bclk, areq0, tHold);
   $width(edge[01,x1]bclk, tWpos); //Width for bclk
   $width(edge[10,x0]bnres, tWneg); //Width for Reset
endspecify
```

The `specparam` declares parameters inside specific blocks. The above example assigns one set of "minimum:typical:maximum" delays for rising transitions and another set for falling transitions from `bclk` to `agnt0` signals.

The setup and hold times are defined for the positive edges of `bclk` and `areq0` using `$setup` and `$hold` system tasks. The `$width` system task specifies the duration of the signal levels from one clock edge to the opposite clock edge. It reports a violation if the interval from the falling `edge(10,x0)` of `bnres` to `rising edge(01,x1)` is less than `tWneg`.

The system-timing check determines the time between the two events and compares the elapsed time to the specified minimum or maximum time limits. The timing violation is reported if the elapsed time occurs outside the specified time limits.

3.3.3 Decoder

The decoder decodes the transfer addresses and selects slaves on the bus appropriately. The decoder in the design can select 16 slaves with 16MBytes of address range for each slave. It generates a bus error for non-selected slave transfer and protected address area transfer. Figure 3-6 shows the signals of the decoder.

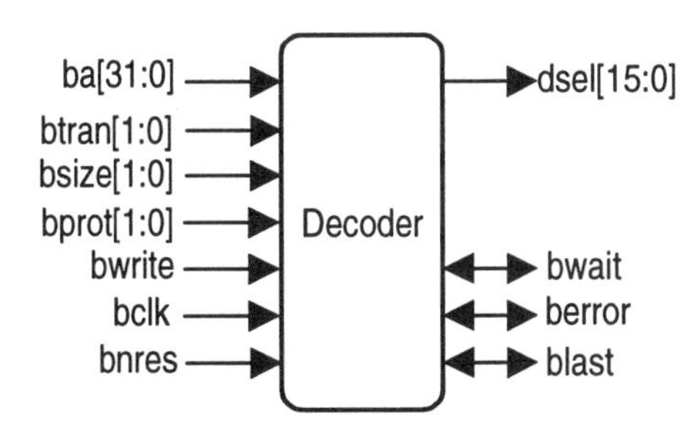

Figure 3-6. Block Diagram of the Decoder

The following signals are connected to the decoder:

- ba—Address bus
- btran—Transfer type indication signals
- bprot—Bus protection control signals
- bwrite—Read/write control signal (0 means read; 1 means write)

- bclk—Bus clock signal
- bnres—Reset signal (active low)
- bsize—Transfer size (8 bits, 16 bits, or 32 bits)
- dsel—Slave select signals
- bwait—Wait signal
- berror—Error signal
- blast—Last transfer signal

3.3.4 ASB Master

A master can initiate bus transfers (read/write) by driving address and control lines. Whenever the bus master wants to perform a data transfer, it generates a request to the arbiter. The arbiter asserts the appropriate grant signal, depending on the priority of the requested master, and the master performs the data transfer. In the Bluetooth SOC, only the processor (ARM7TDMI) is a bus master, and all other peripherals are slaves.

3.3.5 ASB Slave

A slave responds to the request from the master and participates in the transfer. It also responds with error, wait, or successful completion of the data transfer. In the Bluetooth SOC, all the memories, Codec, Bluetooth link controller, USB block, and the ASB/APB bridge are slaves.

3.3.6 ASB/APB Bridge

The ASB/APB bridge is a slave on an ASB. It translates ASB transfers into a suitable format for the slave devices on the advanced peripheral bus (APB). It latches the address, data, and control signals and performs address decoding to generate slave select signals for the peripherals on the APB. Figure 3-7 shows the signals connected to the bridge.

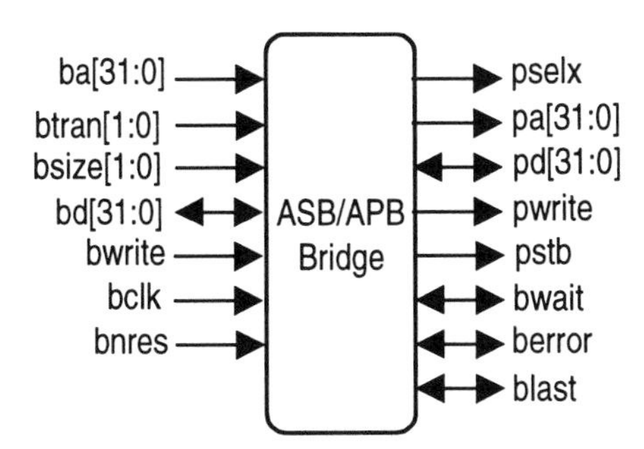

Figure 3-7. Block Diagram of the ASB/APB Bridge

The following signals are connected to the ASB/APB bridge:

- ba—ASB address bus
- btran—ASB transfer type indication signals
- bsize—ASB transfer size (8 bits, 16 bits, or 32bits)
- bd—ASB data bus
- bwrite—ASB read/write control signal (0 means read; 1 means write)
- bclk— ASB bus clock signal
- bnres—ASB reset signal (active low)
- pselx—APB peripheral select signals
- pa— APB address bus
- pd—APB data bus
- pwrite—APB read/write control signal (0 means read; 1 means write)
- pstb—APB peripheral strobe signal
- bwait—ASB wait signal
- berror—ASB error signal
- blast—ASB last transfer signal

3.4 Lint Checking

Lint checking improves the quality of the design code and verification productivity. It is performed on RTL code to analyze and debug code syntax, synthesizability, uninitialized variables, unsupported constructs, and port mismatches. Some of the lint checking tools can do finite-state-machine (FSM) extraction and detect race conditions. It is not efficient to use HDL simulator or synthesis tools to debug a design for these kind of checks. The lint checking tools do not require a testbench to perform checking.

Example 3-6 shows the arbiter block RTL code that is used for lint checking.

Example 3-6. Excerpts from Arbiter Block's RTL Code

```
/***Next State Block****/
always @(current_state or areq1 or areq0 or rid)
 begin
  next_state = idle;
   case (current_state)
    idle:        .....
    master1:     .....
    master2:     .....
   endcase
 end //Next state Block

always @(negedge bclk or negedge bnres)
 begin
  if (bnres == 1'b0)
   begin
    current_state <= idle;
    dset <= 1'b0;
    rid  <= 2'b00;
   end
  else
   current_state <= next_state;
 end
```

When the above code is run with a hypothetical lint checking tool, the tool outputted the following warnings. The warnings are then examined, and the RTL code is fixed accordingly.

```
Processing source file arbiter.v
(W#1)  arbiter.v, line  33: Not all possible cases cov
ered: case (current_state) ... endcase
(W#2)    arbiter.v, line   33: Case statement without
default clause: case (current_state) ... endcase
(W#3)   arbiter.v, line 65: Asynchronous flipflop is
inferred: current_state
(W#3)   arbiter.v, line 66: Asynchronous flipflop is
inferred: dset
(W#3)   arbiter.v, line 67: Asynchronous flipflop is
inferred: rid
```

3.5 Formal Model Checking

In design verification, it is very difficult to detect bugs that depend on a specific sequence of events. Not detecting bugs early can have a serious impact on the design cycle. This has been the main driving force for adopting formal model checking techniques.

Formal model checking uses formal mathematical techniques to verify the behavioral properties of designs. It has been used successfully to verify hardware designs. A model checker explores the entire state space of a design under all possible input conditions, finding bugs that can be difficult to detect through simulation. When a model checker reports a property to be true, the designer can be 100 percent sure that the report is accurate. This is not the case with simulation, unless an exhaustive simulation is performed, which, except for small designs, is impractical.

The formal model checking technique does not require a testbench. The properties to be verified are specified in the form of queries. When the model-checking tool finds an error, it generates a complete trace from the initial state to the state where the specified behavior or property failed. Some model checkers can actually highlight the line in the RTL source code where a signal variable contributing to the error was assigned the incorrect value.

The examples explained in this section were verified with Cadence's FormalCheck® tool.

3.5.1 When to Use Model Checking

Formal model checking provides an exhaustive check of the design properties and can be performed after lint checking the RTL code. This process enables early identification of error conditions that are not obvious candidates for deterministic simulation.

Formal model checking is effective for verifying control-intensive designs, but not for datapath-intensive designs. The designs containing datapaths typically have very large and deep state spaces, and the verification of properties on such designs can be expensive in memory and processor time. However, property-specific reductions can be used to analyze only the part of the design that is relevant to the property and design abstraction.

Arbiters, decoders, FSMs, bus bridges, and other complex control logic blocks are well suited for model checking.

3.5.2 Limitations of Model Checking

Formal model checking does not eliminate the need for simulation but rather supplements it. The model checkers currently available in the industry have capacity restrictions. The size and complexity of the designs that a model checker can handle depend on the property and type of circuit being verified. For certain properties, a model checker might not have the capacity to handle even small designs, whereas for other properties, it could handle larger designs if a reasonable amount of memory and processor resources are available.

Model checking is not yet widely used as a verification tool because earlier model checkers were difficult to run. The property specifications were not intuitive, and designers had to learn special-purpose specification languages. This process has been simplified with the development of easy-to-use graphical user interfaces (GUI).

3.5.3 Model Checking Methodology

Formal model checking requires the RTL code of the design. Most model checkers support Verilog and VHDL descriptions of the design. Some model checkers also require that the RTL contain only the constructs from a synthesizable subset of the language. This ensures that the input to a synthesis tool was verified by a model checker. Figure 3-8 shows the model checking methodology flow.

The model checker reports results when a particular property fails or passes. When a property fails, the model checker outputs a trace from an initial state to the state where the property failed. After debugging, the errors are fixed, the design is recompiled, and verified again. If there are no errors, the RTL code is used for functional simulation.

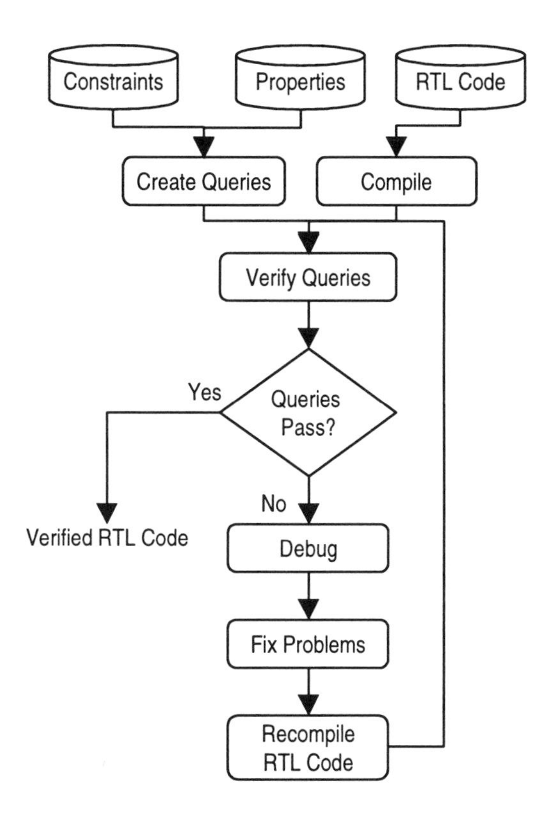

Figure 3-8. Model Checking Methodology Flow

3.5.3.1 Model Checking Properties

Properties specify the behavior to be verified in a model checker. A model checker verifies the following properties of a design.

- **Safety properties**: Specify the behavior corresponding to "something will never happen" or "something always happens." For example, in a bus arbiter, it is never the case where two bus requests are granted simultaneously.

  The model checker verifies that the safety property is satisfied at every state in the state space of the design. In the above example, the property fails if there is at least one state in which two grant signals are true at the same time.

- **Liveness properties**: Describe behaviors that are eventually exhibited. For example, a car waiting at a traffic intersection will eventually get a green signal. For this example, the checker has to verify that in every state where the "car waiting" condition is true, there is a path to a state where the "get a green signal" condition is true.

  The property fails if there is a path from a state where the "car waiting" condition is true to a cycle of states in which the "get a green signal" condition is false. A liveness property cannot be verified by simulation unless all the paths to reach the states in which the eventuality condition is fulfilled have been exercised during the simulation process.

- **Strong liveness properties**: Specify behaviors that involve an enabling condition to be repeatedly true in order for the fulfilling condition to be eventually true.

  The property specification format is as follows:
  -If repeatedly (enabling condition is true)

  -Eventually (fulfilling condition is true)

  A model checker verifies that for every cycle of states in which the enabling condition is true, there is a path to a state where the fulfilling condition is true. A property failure is indicated by a cycle of states in which the enabling condition is true, but there are no paths to any state in which the fulfilling condition is true.

Various model checkers use a specification language called computation tree logic (CTL) to specify properties. A CTL formula for a property consists of Boolean expressions of the state variables in the design and temporal operators. A temporal operator has two parts:

-A path quantifier:
  A—Indicates all paths
  E—Indicates a path exists

-Temporal modalities:
  G—Global operator

F—Eventual operator
X—Next state operator
U—Until operator

For example, the following property specifies that from each design state where the `req` signal is true, it can always reach a state where the `grant` signal is true. In other words, a request is always followed by a grant. When a model checker verifies this property, it explores all paths from each state where the `req` signal is true to determine whether or not there is a path leading to a cycle of states in which the `grant` signal is not true. Verifying this property requires an exhaustive simulation.

```
-AG(req -> AF(grant))
```

In the following example, the property specifies that, at any state of the design if the `reset` signal is true, there is a path from that state to the initial state. In other words, starting from any state, the design can be initialized by using the `reset` signal.

```
-AG (reset -> EF (initial_state))
```

3.5.3.2 Model Checking Constraints

Constraints describe the design environment for verification purposes. Constraints and properties are complementary. A constraint on the inputs of a module can be verified as a property on the outputs of the module driving those inputs. A model checker verifies a property subject to any associated constraints. Without constraints, a model checker might report a property failure due to conditions on the inputs that might never be true. For this reason, generate the constraints under which a corresponding property needs to be verified before running a model checker.

For designs that cannot be handled by a model checker due to capacity limitations, the state space to be explored can be restricted by specifying the constraints on design inputs. This might enable successful verification of the property. However, use caution when specifying the constraints, because too many restrictive constraints can create an empty state space to be explored by the model checker. Also, be sure that the constraints are consistent. Inconsistent constraints might be true individually, but when grouped together, inconsistent constraints can cause an empty state space. This might result in no errors being reported.

Model checkers can perform a number of checks to find common instances of an over-constrained design. They provide options to locate constraints that are mutually inconsistent. They usually report the number of states traversed when verifying a property and the size of the state space for the design. These numbers indicate whether or not the design is over-constrained.

The only conclusive way to determine whether the design is over-constrained is to do the following:

- Convert all constraints on inputs to properties on outputs that drive these inputs
- Verify the output properties

Constraints are an integral part of a bottom-up hierarchical verification process. As verification moves up the design hierarchy, constraints at lower levels become properties to be verified at the next level in the hierarchy. Constraints can be used to automatically generate random simulations. Random stimuli that are compliant with the constraints are generated for simulation. Assertion checks in simulation corresponding to the constraints are used to monitor whether the constraints are violated.

The following two types of constraints are available:

- **Safety**: Similar to the safety properties.
- **Fairness**: Complements of the liveness properties. They are often used to specify exclusion of unwanted behavior from the verification exercise.

  For instance, the following CTL formula specifies a constraint for the bus arbiter example:

  AG ((req ==1) ->(req == 1) **U** (grant == 1))

  It states that it is always the case that once the condition `req==1` is true, it will remain true until the condition `grant==1` becomes true. In this constraint, any state transition corresponding to the condition `req!= 1` (the complement of `req = 1`) is not explored by the model checker.

3.5.4 Performing Model Checking

Model checking involves the following methodology steps.

1. **Extract properties**

   Formal specification of hardware design is typically not available. Usually this information is obtained from an informal description in the design's functional specification, design documents, and informal communication with the designer. This information can be used to derive manually the properties to be verified. The information is then translated to the specification format of the model checker, such as CTL.

2. **Partition the design**

   Model checkers usually have a limited capacity. Because a model checker explores the design's entire state space, state explosion becomes a problem, preventing the design as a whole to be verified. One solution is to partition the design into smaller parts and verify each part separately.

   Because most designs are done in a modular fashion or use a hierarchical design methodology for model checking purposes, the same partitioning used by the designer can be used here. However, verifying a part might require using an abstract model of the other parts. While creating the abstract model, ensure that the original functionality is not lost in the abstraction process. For each part to be verified independently, consult designers to determine properties and constraints.

3. **Model the environment**

   A design is verified in conjunction with its environment, which imposes restrictions on the feasible behavior of the design. If environmental restrictions are not applied, false negatives can occur due to design behavior resulting from unlikely inputs. All possible interactions of the design with its environment must be captured from the design specification in the form of constraints. The constraints are specified using the same mechanism for specifying properties in a model checker.

4. **Debug the RTL**

   When the verification of a property fails in a model checker, it outputs a counterexample, which is usually the trace of the states in the shortest path from an initial state to the state at which the incorrect value was assigned. The cause of the error might not be due to a design error, but to an incorrect property specification, abstraction of the environment, or reduction step.

   It is important to thoroughly analyze the error and pinpoint the exact cause. In some model checkers, a counterexample is output as a timing diagram, which is viewed using a waveform viewer. The timing diagram shows the incorrectly assigned signals and other signals that are part of the property being verified. It is possible to return to the source code line in which the incorrect assignment was made. This information can be used to backtrack through the source code and locate the error. The counterexample is also used to produce a simulation test that can be input to a simulator. This test can be used to reconstruct the error in the design environment and also to verify whether the error has been corrected after a fix to the source RTL code.

The tool FormalCheck is used to verify the arbiter, ASB/APB bridge, and decoder blocks in the Bluetooth SOC design. FormalCheck supports a synthesizable subset of Verilog and VHDL designs. The properties to be verified are in the form of queries, which consist of a set of properties and constraints. These queries can be verified either in batch mode or individually.

When a property fails, FormalCheck displays the counterexample as a waveform diagram. It locates the line in the RTL source code where a signal variable is assigned a value corresponding to the point on the waveform. The waveforms can be viewed by using the waveform viewer associated with the tool. The line in the source code that caused the identified signal state is highlighted in the source window of the FormalCheck user interface.

The following sections demonstrate how formal model checking is performed on the arbiter, ASB/APB bridge, and decoder blocks used in the Bluetooth SOC design.

3.5.4.1 Constraint Definitions

This section describes all the constraints used in model checking the arbiter, ASB/APB bridge, and decoder blocks.

- The areq0_clk, areq1_clk, areq2_clk, and areq3_clk request signals for the arbiter are defined to be periodic.
- bclk—50 percent duty cycle clock that starts low and remains low for one time unit.
- bclk signal—Has two units of low phase and one unit of high phase: bclk_a, bclk_2_1.
- bnres—Reset signal starting high and transitioning to low at 10 time units.
- Reset signal with a constant value of 1: bnres_a, bnres_1.
- btran[1:0]—Defined to start low and stay low for seven time units, beyond which the signal remains high forever.
- ba[31:26] signal constraints—Defined as periodic in the decoder verification, whereas in the APB bridge verification, they are defined with only one high-to-low or low-to-high transition: ba31, ba30, ba29, ba28, ba27, ba26.
- ba[30]—Defined to start low and stay low for one time unit, beyond which the signal remains high forever: ba30_0.

- dselx—Defined to start low and remain low for two time units, transition to high and remain high for two time units, beyond which the signal remains low forever.

3.5.4.2 State Variable Definitions

This section describes the state variables used in the examples.

`Clock_tick_count` obtains the number of clock ticks elapsed between two events in the ASB/APB bridge model checking. It is defined as follows:

```
if (bridge.pstb == 1 && bridge.bclk == rising) then
Clock_tick_count = Clock_tick_count + 1
else if (bridge.pstb == 0) then
Clock_tick_count = 0
```

3.5.4.3 Model Checking the Arbiter

The arbiter ensures that only one bus master at a time is allowed to initiate data transfers. The arbitration scheme is not enforced, so the "highest priority" or "fair" algorithms can be implemented, depending on the application requirements.

The following properties are verified for the arbiter:

- **Mutual Exclusion**: At any given time only one of the four 'agntx' can be active. This property verifies that at any time no more than one master is granted the bus. Table 3-1 gives the mutual exclusion property details.

Table 3-1. Mutual Exclusion Property of the Arbiter

Property Specification	Constraints	State Variables	Results
Never ((arbiter.agnt0 + arbiter.agnt1 + arbiter.agnt2 + arbiter.agnt3) > 1)	bclk, bnres	None	Fail

Figure 3-9 shows that two grant signals, agnt0 and agnt1, are active at the same time, indicating an error.

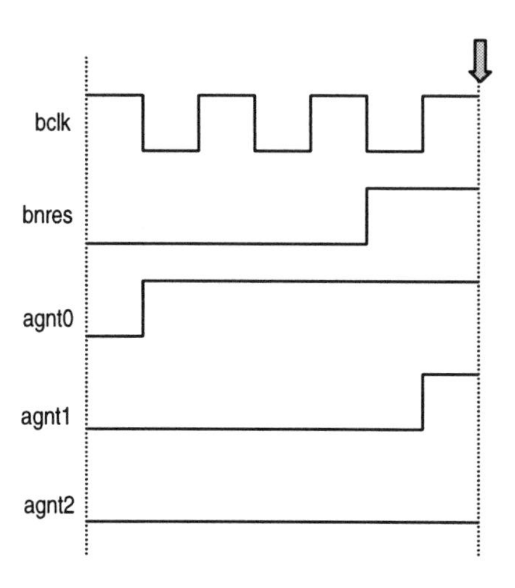

Figure 3-9. Failed Mutual Exclusion Test Waveforms

- **Liveness**: Every bus request eventually gets a grant. This property verifies that all bus requests are eventually serviced. Table 3-2 gives the property details.

Table 3-2. Liveness Property of Arbiter

Property Specifications	Constraints	State Variables	Results
After (arbiter.areq0 == 1) Eventually (arbiter.agnt0 == 1)	bclk, bnres_1, areq0_clk, areq1_clk, areq2_clk, areq3_clk	None	Fail

Figure 3-10 shows that the grant signal, agnt0, is granted for bus request, areq0.

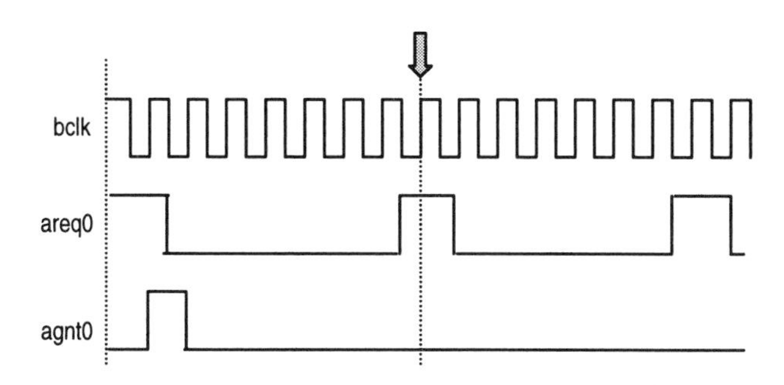

Figure 3-10. Waveform of Failed Liveness Test

3.5.4.4 Model Checking the ASB/APB Bridge

The ASB/APB bridge converts ASB transfers into a suitable format for the slave devices on the APB. The bridge provides latching of all address, data, and control signals, as well as a second level of decoding to generate slave-select signals for the APB components.

The following properties are verified for the APB bridge.

- `pstb` is a peripheral strobe line and should be active for only one clock cycle. Table 3-3 shows the details of the `pstb` signal property.

Table 3-3. pstb Signal Property Details

Property Specifications	Constraints	State Variables	Results
After (bridge.pstb == 1 && bridge.bclk == rising) Never (Clock_tick_count > 1)	bclk_2_1, bnres_1	Clock_tick_count	Pass

- The falling edge of pstb is aligned with the falling edge of bclk. Table 3-4 shows the property details.

Table 3-4. pstb Falling Edge Property Details

Property Specifications	Constraints	State Variables	Results
After (bridge.pstb == 1 && bridge.bclk == rising) Never (Clock_tick_count > 1)	bclk_2_1, bnres_1	Clock_tick_count	Pass
After (bridge.pstb == 1 && bridge.bclk == rising) Never (bridge.pstb == 0) Unless After (decoder.bclk == falling)	bclk_2_1, bnres_1	None	Pass
After (bridge.pstb == falling) Never (bridge.bclk == falling) Unless After (bridge.bclk == 1 && bridge.pstb == 0)	bclk_2_1, bnres_1	None	Pass
After (bridge.bclk == rising) Never (bridge.pstb == rising) Unless After (bridge.bclk == 0)	bclk_2_1, bnres_1	None	Pass

- pselx is a peripheral select line and should be stable when pstb is active. Table 3-5 shows the corresponding property details.

Table 3-5. pselx Signal Line Property with Reference to pstb Line

Property Specifications	Constraints	State Variables	Results
After (bridge.pstb == rising) Never (bridge.pselx[x] == rising || bridge.pselx[x] == falling) Unless After (bridge.pstb == falling)	bclk, bnres_1, ba31, ba30_0, dselx	None	Pass

- When pselx is the active peripheral address, pa[31:2] should remain stable. (APB peripherals are accessed on word boundaries, and the lowest bits of the address bus, pa[1:0], are not usually required.) Table 3-6 shows the property details.

Table 3-6. pselx Signal Line Property with Reference to Address Lines

Property Specification	Constraints	State Variables	Results
After (bridge.pselx[0] == 1) Never (bridge.pa[31] == rising || bridge.pa[31] == falling) Unless After (bridge.pselx[0] == 0)	bclk, bnres_1, ba31, ba30_0, dselx	None	Fail

Figure 3-11 indicates the instability of pa[31], pa[30], and pa[29] when psel[0] is active.

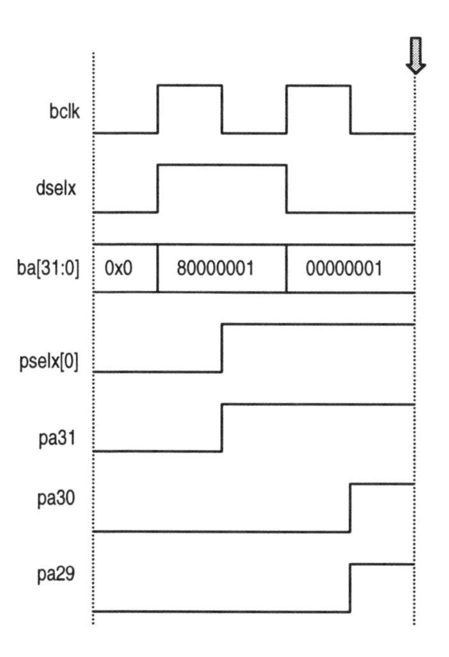

Figure 3-11. Waveform of Failed Signal Stability Test

3.5.4.5 Model Checking the Decoder

The decoder performs the decoding of transfer addresses and selects slaves accordingly. It ensures that the bus remains operational when no bus transfers are required. Refer to Section 3.3.3 for details on the decoder.

The following properties are verified for the decoder.

- **Mutual Exclusion**: At any given time only one of the 16 `dselx` can be active. This property verifies that no more than one slave device is active at a time. Table 3-7 shows the property details.

Table 3-7. Mutual Exclusion Property of the Decoder

Property Specifications	Constraints	State Variables	Result s
Never ((decoder.dsel[15] + decoder.dsel[14] + decoder.dsel[13] + decoder.dsel[12] + decoder.dsel[11] + decoder.dsel[10] + decoder.dsel[9] + decoder.dsel[8] + decoder.dsel[7] + decoder.dsel[6] + decoder.dsel[5] + decoder.dsel[4] + decoder.dsel[3] + decoder.dsel[2] + decoder.dsel[1] + decoder.dsel[0]) > 1)	None	None	Pass

- For an address-only transfer, the decoder responds with `done` and no slaves are selected. An address-only transfer is characterized by `btran[1:0] = 00`. Table 3-8 shows the property details.

Table 3-8. Address-only Property Details of the Decoder

Property Specifications	Constraints	State Variables	Result s
After (decoder.bclk == falling) always ((((decoder.dsel[15] + decoder.dsel[14] + decoder.dsel[13] + decoder.dsel[12] + decoder.dsel[11] + decoder.dsel[10] + decoder.dsel[9] + decoder.dsel[8] + decoder.dsel[7] + decoder.dsel[6] + decoder.dsel[5] + decoder.dsel[4] + decoder.dsel[3] + decoder.dsel[2] + decoder.dsel[1] + decoder.dsel[0]) == 0) && (decoder.bwait/ Output == 0 && decoder.berror/Output == 0 && decoder.blast/Output == 0))) Unless After ((decoder.btran[1] == 1 || decoder.btran[0] == 1) || (decoder.bnres == 0))	bclk bnres btran0 btran1	None	Pass

- `dselx` should change only during the high phase of `bclk`. `dselx` can never change during the low phase of `bclk`. Table 3-9 shows the property details.

Table 3-9. dselx Property of the Decoder

Property Specifications	Constraints	State Variables	Results
After (decoder.bclk == falling) Never ((decoder.dsel[x] == rising || decoder.dsel[x] == falling)) Unless (decoder.bclk == rising)	bclk bnres	None	Pass

- Verify address mapping according to the SOC address map. This ensures that the correct `dselx` signal is active when the corresponding address is present on address bus `ba`, as per the Bluetooth SOC memory mapping table. This property, combined with mutual exclusion, guarantees that only the correct slave device is selected for transactions. Table 3-10 shows the property details.

Table 3-10. dselx Signal Property of the Decoder

<table>
<tr><th>Property Specifications</th><th>Constraints</th><th>State Variables</th><th>Result s</th></tr>
<tr><td>After (decoder.btran[1] == 1 && decoder.btran[0] == 0)
always (decoder.dselx[12] == 1)
Unless (decoder.btran[1] == 0 && decoder.btran[0] == 0) || (decoder.ba[31] != 1 || decoder.ba[30] != 0 || decoder.ba[29] != 1 || decoder.ba[28] != 1 || decoder.ba[27] != 1 || decoder.ba[26] != 0))</td><td>bclk bnres_a btran1_1 ba31, ba30, ba29, ba28, ba27, ba26</td><td>None</td><td>Pass</td></tr>
</table>

- For a non-sequential transfer, the decoder should assert a `bwait` signal to allow address decoding. Table 3-11 shows the property details.

Table 3-11. Non-sequential Transfer Property of the Decoder

Property Specifications	Constraints	State Variables	Results
After (decoder.btran[1] == falling) always (decoder.bwait_r == 1) Unless (decoder.bclk == rising)	bclk, bnres	None	Pass

3.6 Functional Verification/Simulation

Functional verification tests the functionality of the DUT using the testbench. The testbench is created based on the specifications of the design. There are mainly three functional verification approaches: black-box, white-box, and gray-box. The functional verification is performed using event-based and cycle-based simulators.

3.6.1 Black-Box Verification Approach

In this approach, the design is treated as a black box, and the internal design details are unknown for verification. The testbench is created based on the block specification. The errors in the design can be detected only at the output, since the approach does not provide insight into the design details. To stimulate the errors, exhaustive test vectors need to be authored.

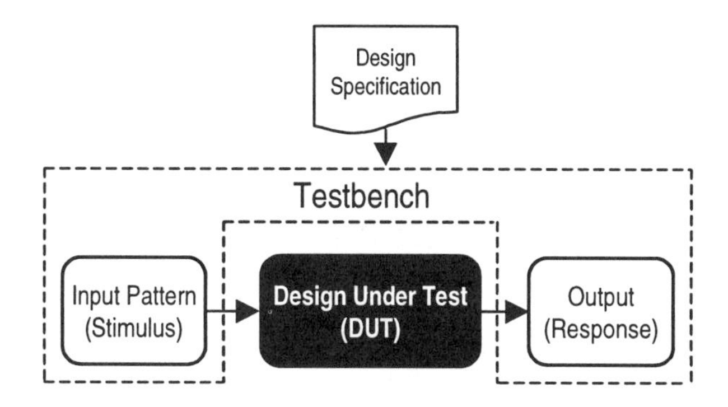

Figure 3-12. Black-Box Verification Approach

The black-box approach focuses on the functional requirements of the design. It attempts to find the following types of errors:

- Initialization and termination errors
- Interface errors
- Performance errors
- Incorrect or missing functions

The black-box approach provides poor observability and controllability, making the debugging task very difficult. It can ensure that the design functions as expected for the given input stimuli, but it does not ensure that the input stimuli fully exercises the design code.

3.6.2 White-Box Verification Approach

This approach provides good observability and controllability for verification. It is also called as structural verification. As shown in Figure 3-13, the design data and structure are visible for verification. The stimulus for corner cases can be easily generated, enabling the source of errors to be detected and identified.This approach is widely used for verification in design houses.

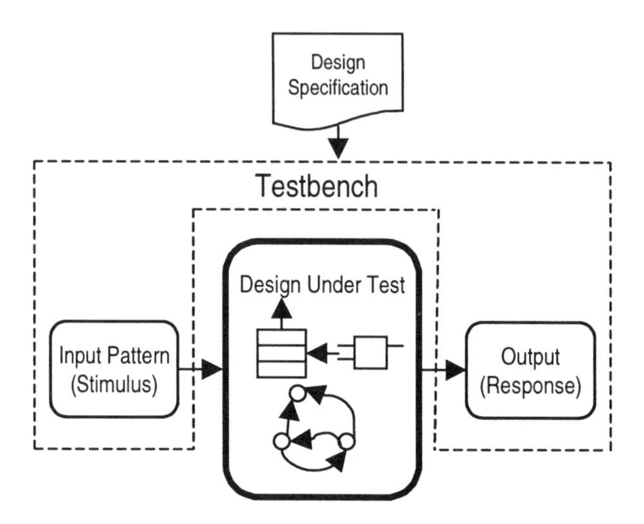

Figure 3-13. White-Box Verification Approach

3.6.3 Gray-Box Verification Approach

In this approach, some of the details of the DUT are known, but not all of the relevant ones to the function that is being verified. This may be because of contractual restrictions, or because the user does not want to verify at greater level of detail. This approach is a mix between white-box and black-box verification.

3.6.4 Simulation

Functional verification/simulation uses event-based or cycle-based simulators. An event-based simulator operates by taking events one at a time and propagating them through a design until a steady state condition is achieved. The design models include timing and functionality. Any change in input stimulus is identified as an event and will be propagated through each stage in the design. A design element may be evaluated several times in a single cycle due to the different arrival times of the inputs and to the feedback of signals from downstream design elements. While this provides a highly accurate simulation environment, the speed of the execution depends on the size of the design and can be relatively slow for large designs.

Cycle-based simulators take a different approach. They have no notion of time and evaluate the logic between state elements and/or ports in the single shot. Because each logic element is evaluated only once per cycle, this significantly speeds the execution time; however the simpler model used by the cycle-based simulator (no timing, fewer logic states) can lead to simulation errors. Cycle-based simulations also put restrictions on the designs that they can handle, for example, they only function on synchronous logic.

3.7 Protocol Checking

Protocol checking verifies that no bus protocol violations or block-to-block interconnect violations occurred during simulation. The protocol-checking exercises bus-centric tests; it verifies correct bus operation while applying stimulus through stimulus generators. Figure 3-14 shows the block diagram of block protocol checking. The protocol checking ensures that a block (IP) can be integrated into a system and is compliant to the interconnection bus.

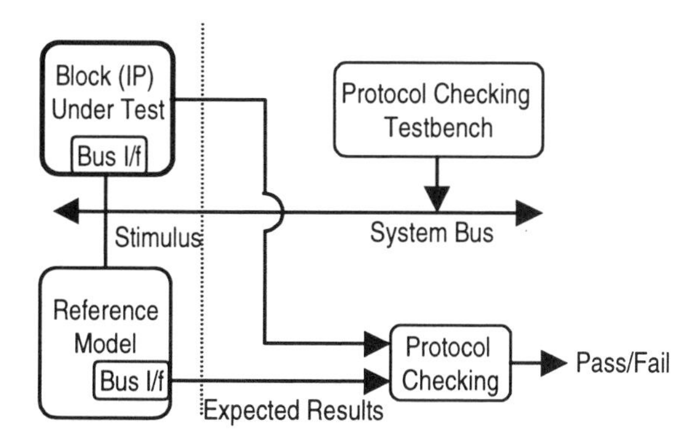

Figure 3-14. Protocol Checking

The bus protocol checkers can either be self-contained, capturing the expected behavior of transactions on the bus, or be based around a reference model. In the latter case, the stimulus received by the block under test would be driven into the reference model as well, and the resulting output from the two models is compared.

A protocol-checking testbench generates bus cycles on the system bus. This testbench is created using write, read, burst write, and burst read routines. These routines emulate the external memory access and the address and data bus interface by using their processor controls.

3.7.1 Memory/Register Access Signals

The following list of signals are used during a memory and register access.

- Address bus (32-bits)—Address bus retains the old value until it changes.
- Data bus (32-bits)—Bidirectional bus that is tri-stated whenever data is not valid.
- Chip select—Active low signal that has the same setup and holds values as that of the address bus. This signal toggles on individual reads and writes, but it remains low in burst mode cycle.
- Read/Write control—0 for read and 1 for write.
- Clock—Input clock to the processor (bclk).
- Interrupts—If an interrupt occurs during a burst operation, the chip select is de-asserted and the interrupt serviced. The individual read or write is completed, and the interrupt is serviced before returning to complete the burst operation.

  If the interrupt occurs during an individual read or write operation, the current operation is completed before the interrupt is serviced. After completing the current instruction (read or write), the read control, write control, and chip select signals are de-asserted before exercising the interrupts.

3.7.2 Protocol Checking Examples

The protocol checkers are implemented for each of the Bluetooth SOC blocks, including the ASB master, ASB slave, arbiter, and decoder. They flag any bus violations with an appropriate message. These messages localize an error and help identify the source of the bug. The following section provides examples of protocol checking.

3.7.2.1 Processor Routines

The following routines are used with the processor (ARM7TDMI).

The `write` routine writes to a register or a single location in memory. The inputs are address and data. Example 3-7 shows the code.

Example 3-7. `write` Routine

```
// Task/Function Name: write

// Synopsis: write (addr, data)
// Description:
// addr : Address of memory or register location to
// write into.
// data : The data that has to be written into this
// location.

     task write_arm;
     input [31:0]  addr;
     input [31:0]  data;

     begin
      `ifdef arm_model
       `ifdef thumb
// execute instructions for memory transfers in the thumb
//mode
       `else
// execute instructions for memory transfers in the regular
// mode
       `endif
      `else
       begin
         nreq  = 1'b0;
         seqq  = 1'b0;
         nrw   = 1'b0;
         nwait = 1'b1;
         ape   = 1'b0;
         dbe   = 1'b1;
         @ (negedge blck)
```

```
        fork
          {nmreq, seq} = #('Tmsh) 2'b01;
          nwait        = #('Tmclkp/2.0 - Tws) 1'b0;
          nrw          = #('Tape) 1'b1;
          a            = #('Tape) addr;
          dbe          = 1'b0;
        join
       @ (negedge bclk)
        fork
          ape         = #('Taph) 1'b1;
          nwait       = #('Twh) 1'b1;
          dbe         = 1'b1;
          d_out       = #('Tdout) data;
        join
     end
    'endif
   end
   endendtask //end write
```

The `write-burst` routine writes to successive locations in memory. Inputs to this routine are an address, the number of locations to be written, and data. To conform with the write routine, add the offset and wait-states. If the lowest bit of the data is "z," a random number is generated and written to the memory location, and a copy is maintained in the local images of the memories. Otherwise, the data supplied to this routine is used.

The `read` routine reads from a register or a single location in memory and uses address and compare data as its inputs. If the write task was created with wait-states and address offsets, those same items must be used in this routine. If the lowest bit of data is a "z," the copy of the random data written during a write operation is compared against data on the data bus. Otherwise, the data supplied to this task is used for comparison. When this routine reads random data where the lowest bit of the compare is "z," the compare data is obtained from local images of the memory if the address is a memory location. Otherwise, the data is stored in `data_random_reg` when a random write is used. Example 3-8 shows the code.

Example 3-8. `read` Routine

```
// Task/Function Name: read
// Synopsis: read (addr, data)
```

```
// Description:
// addr : Address of memory or register location to read
// from.
// data : The data that has to be used for comparison.

      task read_arm;
      input [31:0]    addr;
      input [31:0]    data;
      begin
       `ifdef arm_model
        `ifdef thumb
// execute instructions for memory transfers in the thumb
// mode
        `else
// execute instructions for memory transfers in the regular
// mode
        `endif
       `else
         begin
         nmreq  = 1'b0;
         seq    = 1'b0;
         nrw    = 1'b0;
         nwait  = 1'b1;
         ape    = 1'b1;
         dbe    = 1'b1;
         a      = addr;
         @ (negedge bclk)
          fork
            {nmreq, seq} = #(`Tmsh) 2'b01;
            nwait        = #(`Tmclkp/2.0 - Tws) 1'b0;
            dbe          = 1'b0;
          join
         @ (negedge bclk)
          fork
            ape        = #(`Tmclkp - `Taps) 1'b0;
            nwait      = #(`Twh) 1'b1;
            dbe        = 1'b1;
          join
         @ (negedge bclk)
```

```
          check_data_arm (data);
        end
     'endif
    end
    endendtask //end read
```

The `read-burst` routine reads from successive locations in memory. This routine uses the address and number of locations as its inputs. The data on the bus is compared against the data that is stored locally in the images of the memories.

The `check_data` routine compares the data captured on the bus with the expected data and reports any violations. Example 3-9 shows the code.

Example 3-9. `check_data` Routine

```
// Task/Function Name: check_data_***
// Synopsis : check_data_arm (compare_data)
// Description :
task check_data_arm;
input [31:0] compare_data;
begin
  'ifdef arm_model
    'ifdef thumb
//execute instructions for memory transfers in the thumb
// mode
    'else
//execute instructions for memory transfers in the reg
//ular mode
    'endif
   'else
      if (d_in !== compare_data)
        begin
          $display("Data violation at %d ns", $time);
          $display("At address %h, compare data is %h and
          %h was read" , a, compare_data, d_in);
        end
  'endif
end
endtask //end check_data_arm
```

3.7.2.2 ASB Master

The master functionality is a subset of the full, required ASB master functionality. In wait state, `{bwait,berror,blast} = 3'b100`, the address signals, `ba`, or the transfer type signal, `btran`, must not be changed. The bus protocol checker must flag any violations in these conditions with a message.

Example 3-10 shows Verilog code for protocol checks of the ASB master under the condition that "address must not change during wait condition."

Example 3-10. Protocol Check for the ASB Master

```
reg [31:0] prev_address;
reg [1:0] prev_tran;
always @(bclk)  // AMBA bus works on both edges of the clock.
 begin
   prev_address = ba:
   prev_tran = btran;
 end
 always @(bwait)
  begin
// Checking for Wait state
    if ({bwait, berror, blast} == 3'b100)
     begin
// comparing with the previous address
      if (prev_address != ba)
       $write("ERROR:Master-- Address changed during Wait
       state @time %d Address =%h\n", $time, ba);
// comparing with the previous address type signal.
      if (prev_tran != btran)
       $write("ERROR:Master-- Address type changed during
       Wait state @time %d btran =%b\n", $time, btran);
    end
end
```

3.7.2.3 Decoder

The decoder functionality is a subset of the full required decoder functionality. Following are the protocol requirements.

- During the reset state, no slave must be selected. The transfer response signals, `bwait`, `blast`, and `berror`, must be deasserted.
- Two slaves cannot be selected at one time.
- If the address, `BA`, is out of range and the transfer type is not addressed, the error signal `berror` must be asserted.

The following examples of Verilog HDL code show protocol checks for the decoder.

Example 3-11. Checking Condition "During Reset All Transfer Response Signals Must Be Low"

```
always @(bnres or bwait or berror or blast)
    if (!bnres)    // Active low reset
    begin
// Assuming synchronous reset operation
      @(negedge bclk)
      #('strobe_time)      // Settling time.
      if ({bwait,blast,berror}!= 3'b000)
      begin
// Detailed message gives information regarding the nature
// of the error.
        $write (ERROR:Decode-- Transfer response signals not
        deasserted during reset at time:%d Expected: 000
        Got:%b\n, $time, {bwait,blast, berror});
      end
    end
```

Example 3-12. Checking Condition "At Any Time Only One or None of the Slaves Must Be Selected"

```
integer slave_value, i;
     always @('slave_signals)
     // slave signals are dsel0 or dsel1 or dsel2...
     begin
       slave_value = 0;
       for (i =0; i < 'number_of_slaves; i = i + 1)
```

```
    slave_value = slave_value + ({'slave_signals} >> 1);
    // slave_signals is {dsel0,dsel1...}
    if ((slave_value > 1) || (slave_value === 1'bx))
      $write(ERROR:Decode-- More than one slave signal
      selected or one of the slaves driven to x @timed
      Got:%b\n, $time, {'slave_signals});
  end
```

Example 3-13. Checking Condition "berror Signal Is Flagged When the Address Is Out of Range"

```
always @(btran or ba)
begin
   @(negedge bclk) // Wait for the negedge of Clock
  #('strobe_time);  // Wait for a strobe time.
  if ((ba != address_out_of_range) && !berror && (btran !=
      'atran))
    $write("ERROR:Decode-- Error not asserted when address
      out of range @time%d Address = %h Got:%b\n", $time, ba,
      berror);
 end
```

3.8 Directed Random Testing

The quality of a functional verification environment depends on the stimulus that is applied to a DUT. An exhaustive test vector set can be written using all combinations of the input signals, but this is not feasible, since it increases the simulation time tremendously. In directed random testing, random address, data, and control signals are driven onto a bus, and one or more bus protocol checkers verify that bus protocol violations do not occur as a result of these cycles. This testing approach is well suited for bus validation.

The testbenches are directed in that the test cycles generated are not purely random but create cycles that stress the design in specific ways. The pattern generators can be set to create specific transaction types, such as read, write, and read-modify-write in a random sequence. For example, 20 percent read, 30 percent write, 50 percent read-modify-write.

Similarly, data and address fields can be generated in a random sequence, but within specified limits or using a limited set of discrete values. These types of tests verify corner conditions and sequential or data-dependent situations that are difficult to identify in simulation. With this methodology, any algorithmic errors are identified and fixed early in the design cycle.

3.8.1 Random Vectors Generation in Verilog and C

Random vectors can be easily generated in Verilog HDL. Verilog HDL provides a list of system tasks to generate the random vectors (see the Verilog language reference manual). The systems tasks of interest are:

- $random
- $dist_uniform
- $dist_exponential
- $dist_poisson
- $dist_t
- $dist_chi_square
- $dist_erlang

The standard C library also has built-in functions to generate random values. For example, the functions are:

- rand
- srand
- rand_r

Example 3-14 shows random vector generation in Verilog HDL. This code assigns random vectors to the signals `ba` (address bus), `bwait`, `bwrite`, and `bprot`.

Example 3-14. Generating Random Vectors in Verilog HDL for Address Bus and Control Signals

```
initial
begin
   BTRAN = 2'b10; // Setting the primary signals.
```

```
    bnres = 1'b1;
    for  ( i = 0; i < 'number_of_vectors; i = i + 1 )
//number of vectors
    begin
    @(negedge bclk);
    #('period - 't_istr)   // Waiting for the setup time.
      if (! bwait)
//The control signals must not be altered when bwait
//is high.
       {ba,bsize,bwrite,bprot} = $random;
//Random values are assigned and it can be checked
//if dsel is selected correctly.
    end
end
```

In Example 3-15, the active low reset is activated for a random number of cycles between 0 and 24. The time between one reset and the next reset is also random.

Example 3-15. Testing the Reset Functionality

```
initial
  begin
   repeat (8)      // The reset is toggled 8 times.
   begin
   bnres = 0;
    j = $dist_uniform(8,2,64);
//The initial seed value is 8 and the values follow
//uniform distribution in the range of 2 to 64.
    #(j * 'period) ;
    #('t_ihnres) bnres = 1;k = {$random} % 1016 + 8;
//Reset is high between 8 to 1023 cycles.
    #(k * 'period);
     end
end
```

In Example 3-16, the code generates a scenario with 70 percent writes and 30 percent reads.

Example 3-16. Testing the Read/Write Functionality

```
initial
begin
    repeat('number_of_transfers)
    begin
    // Generates a random number between 0 to 9.
    j = {$random} %10;        if (j < 7)
     BnWrite = 1;      // 70% of which are Write
    else
     BnWrite = 0;      // 30% are Reads
    if (BnWrite)
      // Write task
        else
      // Read task
    end
end
```

3.9 Code Coverage Analysis

Code coverage involves running simulation on the design while a code coverage tool tracks the number of times each line of RTL code is exercised. The results are analyzed and reviewed by the verification and design teams. The testbench is incorporated with additional tests to simulate lines of code not covered by the initial testbench. Performing coverage analysis on the code increases the confidence level in the design.

3.9.1 Types of Coverage

The coverage analysis types are statement, toggle, state machine, visited state, triggering, branch, expression, path, and signal. Some of the coverage types have alternate names.

3.9.1.1 Statement Coverage

Statement coverage, also called line, block, or segment coverage, shows how many times each statement was executed.

For example, consider the following code:

```
always @ (areq0 or areq1)
 begin
  gnt0 = 0 ;
  if (areq0 == 1) gnt0 = 1 ;
 end
```

The code example contains the following statements:

- First statement: `gnt0 = 0 ;`
- Second statement: `if (areq0 == 1) gnt0 = 1 ;`

The first statement is a single statement. The second statement contains two or three statements. It is considered as two if only `(areq0 = = 1)` is executed, and three if both `(areq0 == 0)` and `(areq0 == 1)` are executed in addition to `(gnt0 = 1)`. The second statement is considered fully covered only if the `areq0 is not equal to 0` condition is tested by the testbench.

3.9.1.2 Toggle Coverage

Toggle coverage shows which signal bits in the design have toggled. This coverage analyzes both RTL and gate-level netlist. It is generally used for gate-level design fault coverage and power analysis.

3.9.1.3 State Machine Coverage

The state machine coverage shows how many transitions of the FSM were processed. It is also called state value, state transition, or FSM arc coverage. Performing state machine coverage ensures that all legal states of the FSM are visited and all the state transitions are exercised.

3.9.1.4 Visited State Coverage

Visited state coverage shows how many states of the FSM were entered or visited during simulation. This coverage analysis is essential for very complex state machines because it finds out whether all the state transitions are visited.

3.9.1.5 Triggering Coverage

Triggering coverage shows whether each process has been uniquely triggered by each signal in its sensitivity list. It is also called event coverage.

For the following code, the triggering coverage reports whether the code is tested for the events on each variable used in the `always @` sensitivity list.

```
always @ (areq0 or areq1 or areq2)
 begin
  ......
 end
```

3.9.1.6 Branch Coverage

Branch coverage shows which case or "if...else" branches were executed. It is also called decision coverage.

For example, for the following case statement, the branch coverage verifies that each of the four cases is covered.

```
case (areq)
0: gnt = 0 ;
1: gnt = 1 ;
2: gnt = 2 ;
3: gnt = 3 ;
end case
```

For an "if ...else" statement, branch coverage checks the `if` statement for both true and false conditions when simulated. The branch coverage even checks whether the `else` statement is not mentioned in the code.

3.9.1.7 Expression Coverage

Expression coverage shows how well a Boolean expression in an `if` condition or assignment has been tested. It is also called condition, condition-decision, or multiple condition coverage.

For example, for the following code, expression coverage reports which combinations of the values for `areq0` and `areq1` are tested in the expression.

```
assign areq = areq0 || areq1 ;
```

In this example, the possible combinations are:

```
areq0  = 0 areq1 = 0
areq0  = 0 areq1 = 1
areq0  = 1 areq1 = 0
areq0  = 1 areq1 = 1
```

3.9.1.8 Path Coverage

Path coverage shows which routes through sequential "if...else" and case constructs have been tested. It is also called predicate or basis path coverage. Path coverage is similar to branch coverage. It handles multiple sequential decisions.

For example, consider the following code:

```
if (areq0) begin
    ......
end
if (areq1) begin
    ......
end
```

The code has four paths. It needs to be evaluated for `areq0 = 0` and `areq1 = 0` and for `areq0 = 1` and `areq1 = 1` for full coverage.

3.9.1.9 Signal Coverage

Signal coverage shows how well state signals or ROM addresses have been tested.

3.9.2 Performing Code Coverage Analysis

Several code coverage tools are available in the industry. This section explains code coverage analysis using Cadence® Coverscan, which provides information about the portions of the Verilog code that have been tested.

Figure 3-15 shows the Cadence Coverscan task flow. It consists of three primary components:

- **Pre-Coverscan**: Pre-processor that annotates the RTL code for recording coverage data. The designer can tailor instrumentation of the code using embedded comments, control files, or command line options.
- **Coverscan Recorder**: Records code coverage data during simulation. It interfaces directly with the HDL simulator through PLI routines.
- **Coverscan Analyzer**: Graphical, post-simulation tool for quickly analyzing coverage results. It can be used to pinpoint coverage problems and change coverage criteria on a module, block, or line basis.

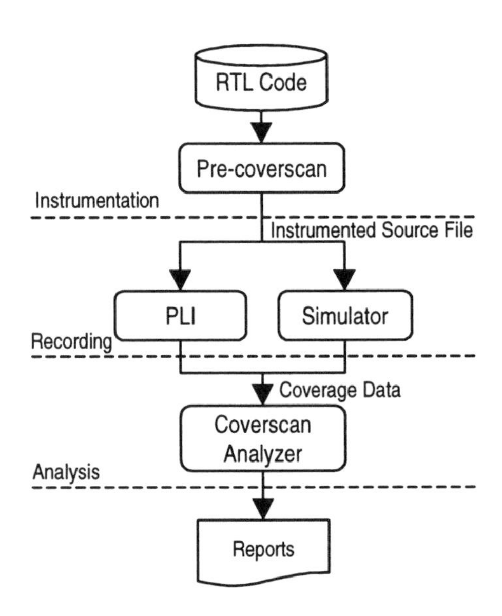

Figure 3-15. Cadence Coverscan Task Flow

The arbiter block in the Bluetooth SOC design is used to illustrate code coverage analysis using Cadence Coverscan. Example 3-2 shows the testbench applied to the arbiter (explained in Section 3.3.1).

Figure 3-16 shows the report after running code coverage on the arbiter block.

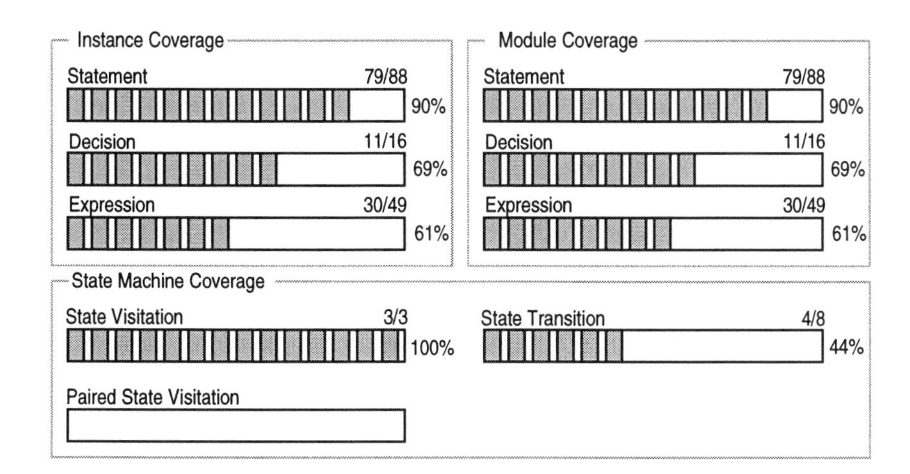

Figure 3-16. Code Coverage Analysis Report

Table 3-12 shows the line and expression coverage. The column labeled S# indicates the number of times a statement in the row is exercised by the testbench. The column labeled E% indicates the expression coverage percentage for the expression in that row. It shows how many times the expression and each of the subexpressions, `(areq1==1)` and `(areq0==0)`, were executed and the results.

Table 3-12. Line and Expression Coverage

<table>
<tr><th>S Cat</th><th>S#</th><th>E Cat</th><th>E%</th><th>Line</th><th>Source: arbiter</th></tr>
<tr><td></td><td></td><td></td><td></td><td>44</td><td></td></tr>
<tr><td></td><td></td><td></td><td></td><td>45</td><td>/***Next State Block***/</td></tr>
<tr><td></td><td></td><td></td><td></td><td>46</td><td>//Synopsys state_vector current_state</td></tr>
<tr><td></td><td></td><td></td><td></td><td>47</td><td>always @(current_state or areq0 or areq1 or rid)</td></tr>
<tr><td></td><td></td><td></td><td></td><td>48</td><td>begin</td></tr>
<tr><td>Executed</td><td>17</td><td></td><td></td><td>49</td><td>next_state = idle;</td></tr>
<tr><td>Executed</td><td>17</td><td></td><td></td><td>50</td><td>case (current_state)</td></tr>
<tr><td></td><td></td><td></td><td></td><td>51</td><td>idle:</td></tr>
<tr><td>Executed</td><td></td><td>Standard</td><td>100</td><td>52</td><td>if (areq1 = = 1 && areq0 = = 0)</td></tr>
<tr><td></td><td></td><td></td><td></td><td></td><td>areq1
| areq0
| | (areq1 == 1)
Count Result | | | (areq0 == 0)
-----+--------+-----+--------------
2 0 0 - 0 -
1 1 1 0 1 1
1 0 1 1 1 0
1 x/z</td></tr>
<tr><td>Executed</td><td>1</td><td></td><td></td><td>53</td><td>next_state = master2 ;</td></tr>
<tr><td>Executed</td><td>4</td><td>Standard</td><td>100</td><td>54</td><td>else if (areq0 == 1 && areq1 == 0)</td></tr>
<tr><td>Executed</td><td>1</td><td></td><td></td><td>55</td><td>next_state = master1 ;</td></tr>
<tr><td>Executed</td><td>3</td><td>Standard</td><td>50</td><td>56</td><td>else if (areq0 ==1 && areq1 ==1 & rid ==2'b00)</td></tr>
<tr><td>Executed</td><td>1</td><td></td><td></td><td>57</td><td>next_state = master1 ;</td></tr>
<tr><td>Executed</td><td>2</td><td>Standard</td><td>25</td><td>58</td><td>else if (areq0 ==1 && areq1 ==1 & rid ==2'b01)
areq0</td></tr>
</table>

Table 3-13 shows the state machine coverage. The column labeled Visit% shows the percentage of the states visited, while the column labeled Transition% shows the percentage of the state transitions in the state machine covered by the testbench.

Table 3-13. State Machine Coverage

State Machine	Visit%	Visit#	Transition%	Transition#	Illegal#
current_state	100	3/3	44	4/9	0

Table 3-14 shows the state coverage for one of the state machines in the arbiter. This view lists each of the states in state machine and the number of times they were visited.

Table 3-14. State Visit Coverage

State	Category	Count
idle (2'h0)	Required	1
master1 (2'h1)	Required	4
master2 (2'h2)	Required	2

Table 3-15 shows the state transition coverage from each of the states.

- An entry of 0 indicates that a required transition did not occur.
- An empty location indicates that an illegal transition did not occur.
- An R indicates that a transition was valid.

Table 3-15. State Transition Coverage

From\To	idle	master1	master2
idle	R 0	R 1	R 0
master1	R 1	R 0	R 2
master2	R 0	R 2	R 0

Summary

Block-level verification is essential when working with SOC designs. More and more, SOCs are using and reusing IPs. These IPs need to be verified early in the design process and during integration using the techniques that are described in this chapter.

References

1. Ellis Richard, Bray Neil, Chaplin David. The armor of IP verification, Integrated System Design, February 2000.

2. Sandler Scott. Debugging, design reuse, and IP integration, Application note, www.chipcenter.com.

3. Thomas Delae. SOC verification based on IP reuse methodology, SAME 99, October 1999.

4. Switzer Scott, Landoll David. Using embedded checkers to solve verification challenges, DesignCon 2000.

5. Keating Michael, Bricaud Pierre. Reuse methodology manual for system-on-a-chip designs, Kluwer Academic Publishers, 1999.

6. Fillipi E, Licciardi A,..... The virtual chip set: A parametric IP library for system on a chip, IEEE Custom Integrated Circuits Conference 1998.

7. Dignam David, Garlick, ... An integrated environment for configurable designs, IEEE Custom Integrated Circuits Conference 1999.

8. Diehl Stan. IP reuse drives SOC design, Portable Design, May 2000.

9. Stadler Manfred, Rower Thomas. Functional verification of intellectual properties (IP): a simulation based solution for an application-specific instruction-set processor, ITC International Test Conference 1999.

10. Saunders Larry. Effective design verification. Integrated System Design, April 1997.

11. Joyce Dan. Code coverage analysis works in hardware design, Integrated System Design, January 1997.

12. Bricaud Pierre J. IP reuse creation for System-on-a-chip design, IEEE Custom Integrated Circuits Conference 1999.

13. Functional verification automation for IP, a whitepaper, www.verisity.com.

14. Anderson Thomas L. The challenge of verifying a synthesizable core, Computer Design, July 1996.

15. Verification navigator data sheet, www.transeda.com.

16. User manuals for Cadence FormalCheck Model Checker, Coverscan, and NC-Sim tools, www.cadence.com.

17. Janick Bergeron. Writing Testbenches : Functional Verification of HDL Models, Kluwer Academic Publishers, 2000.

18. Clarke, E. M, Grumberg Orna, Peled, Doron. Model Checking, MIT Press, 1999.

19. McMillan, Kenneth L. Symbolic Model Checking, Kluwer Academic Publishers, 1993.

20. Bening Lionel, Foster Harry. Principles of Verifiable Rtl Design : A Functional Coding Style Supporting Verification Processes in Verilog, Kluwer Academic Publishers, 2000.

21. Probst, David. Advances in Hardware Design and Verification, Chapman and Hall, 1997.

22. Budkowski Stan, Najm Elie. Formal Description Techniques and Protocol Specification, Testing and Verification, Chapman and Hall, 1998.

CHAPTER 4

Analog/Mixed Signal Simulation

Consumer electronics, such as cellular phones, interactive televisions, automotive subsystems, require analog/mixed signal (AMS) interfaces in addition to major digital blocks. However, embedding AMS blocks in a system-on-a-chip poses a challenge to simulation, since it involves verifying both digital and analog circuits.

This chapter illustrates the following topics:

- Mixed-signal simulation
- Design abstraction levels
- Selecting a simulation environment
- Limitations of current environments
- Using SPICE
- Simulation methodology
- Chip-level verification

Mixed-signal simulation is illustrated with a digital-to-analog converter (DAC) block used in the Bluetooth SOC design.

4.1 Mixed-Signal Simulation

AMS circuits interface to the real world by capturing analog signals and converting them to digital signals, and vice versa. AMS blocks include phase-locked loop (PLL), analog-to-digital converter (ADC), DAC, and others. Today's integrated circuit (IC) technology allow designers to incorporate the AMS blocks along with digital blocks into a SOC, resulting in lower cost, lower power, and enhanced reliability as compared with implementation based on many discrete ICs.

The purpose of analog simulation is to verify that the design under test (DUT) meets the required functionality according to voltage, current, and timing specifications. In digital simulation the purpose is to verify that the DUT meets the required functionality according to predetermined input vectors and timing specifications; voltage precision in the design is not a concern. Figure 4-1 shows the analog and digital circuits and simulation examples.

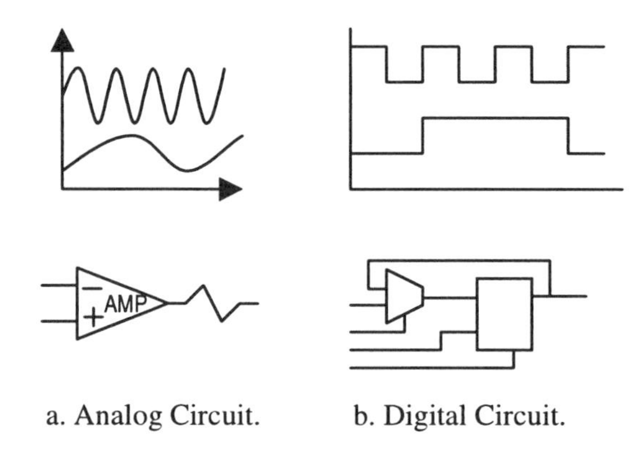

Figure 4-1. Analog and Digital Circuit Simulation

The mixed-signal simulation environment needs to address analog simulation and digital circuit simulation, and the interaction between the two. Currently, most mixed-signal simulation options available couple an existing digital simulator with an analog simulator. Figure 4-2 shows a simple block diagram of the mixed-signal simulation elements.

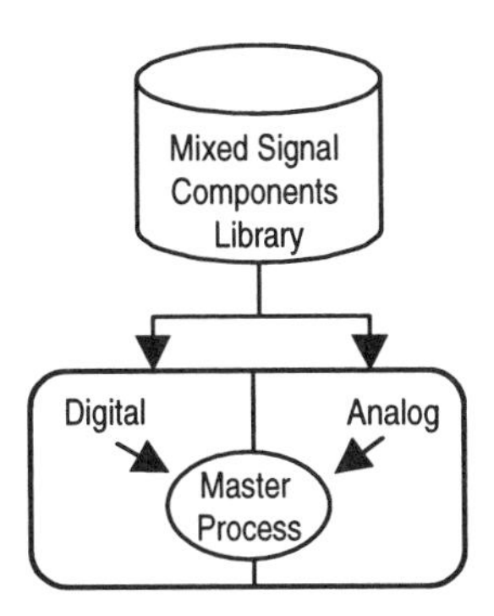

Figure 4-2. Mixed-Signal Simulation Elements

The digital and analog simulators are run as separate processes, with the data transfer between them controlled by a master process. The master process synchronizes the simulators in order to pass the data between them. The following techniques are used to achieve synchronization:

- **Leap frog**: One simulation engine runs ahead of the other, but with a risk that it may have to backtrack if the input is required from the slower simulator. The backtrack may degrade overall simulation speed, depending on the degree and frequency of backtracking.
- **Lock step**: Digital and analog simulations are synchronized at every time step, even if the data transfer between the simulations is not required. The analog simulator determines the step sizes, and the digital simulator uses these values.

4.2 Design Abstraction Levels

Digital and analog designs are described in various top-down abstraction levels for the purpose of design and simulation. Figure 4-3 and 4-4 show the design abstraction levels.

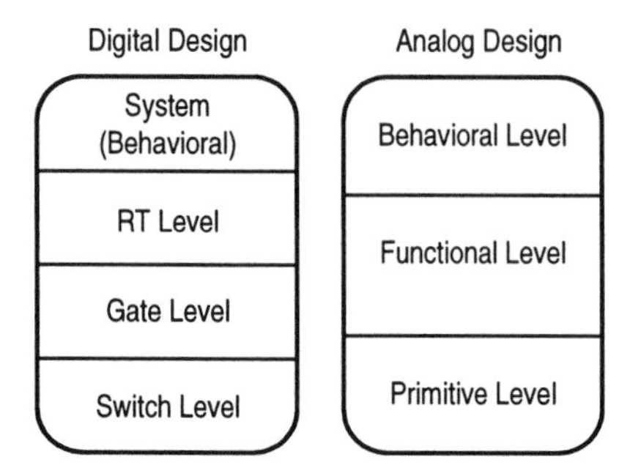

Figure 4-3. Digital and Analog Design Abstraction Levels

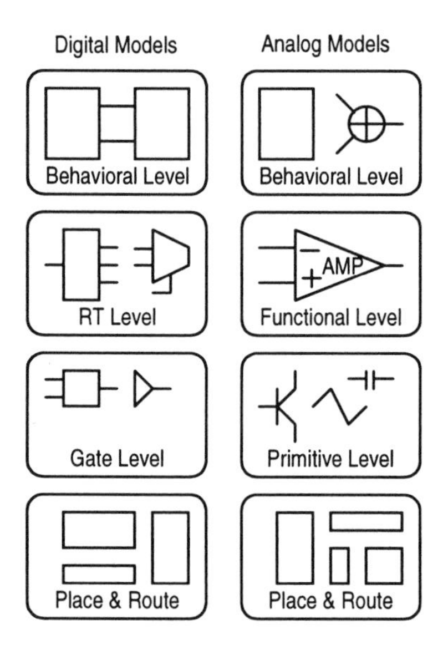

Figure 4-4. Top-Down Abstraction Models

The abstraction levels for a digital design are:

- **System (behavioral) level**: Describes the behavior of the design with few or no details on the structural implementation. This level is used to simulate and prove the basic concepts of the system or design and create structural implementation specifications.

- **Register-transfer level (RTL)**: Describes the function of the design in terms of registers, combinational circuits, buses, and control circuits, with no details on the gate-level implementation. Simulation is done at RTL to verify the logic and timing of the design.
- **Gate level**: Describes the function, timing, and structure of a design in terms of the structural interconnection of logic gates. The logic behavior blocks implement Boolean functions, such as NAND, NOR, NOT, AND, OR, and XOR. The gate level is used to verify the timing of individual signal paths.
- **Switch level**: Describes the interconnection of the transistors that make up a logic circuit. The transistors are modeled as on-off switches. This level is used for verifying timing information more accurately for critical signal paths.

The abstraction levels for an analog design are:

- **Behavioral level**: Describes the behavior of the design with few or no details on the structural implementation. This level is used to simulate and prove the basic concepts of the system or design and create structural implementation specifications.
- **Functional level**: Describes the function of the design with no details on the transistor level implementation. This level is analogous to RTL in digital design.
- **Primitive level**: Describes the operation of a circuit in terms of the voltage-current behaviors of the resistor, capacitor, inductor, and semiconductor circuit components and their interconnection.

IP providers package their IPs with some or all of the above models, along with the appropriate testbenches.

4.3 Simulation Environment

Figure 4-5 shows a typical mixed-signal environment. It consists of the following elements:

- Facility for design entry
- Analog simulator for analog circuit simulation
- Digital/logic simulator for logic simulation

- Link process that manages the communication between analog and digital simulators
- Device and interface element model library
- Simulation testbench
- Facility to display output waveforms and results

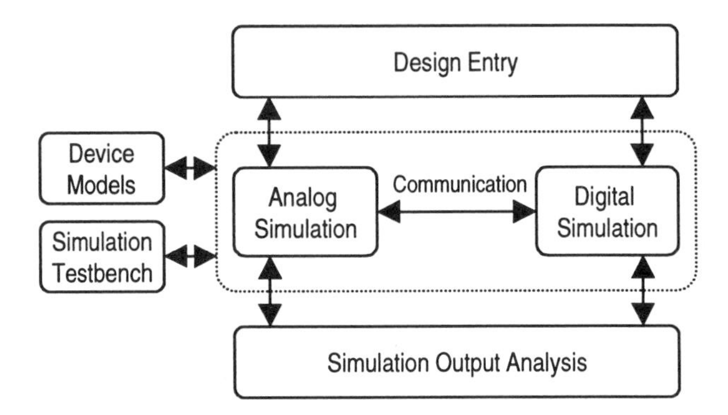

Figure 4-5. Mixed-Signal Simulation Environment

4.3.1 Selecting a Simulation Environment

Following are some of the parameters to considered when selecting a mixed-signal simulation environment.

- **Performance and accuracy**: The overall speed and accuracy of the mixed-signal environment is determined by the analog simulator used. Performance depends on design capacity, clock speed, timing constraints, simulation set-up time, and the synchronization technique used.
- **Library models**: Check the availability of all the critical library models required for the design. Also check the accuracy and performance of the models with real devices. The environment should allow interfacing models at different abstraction levels for the same design.
- **Interface models**: The digital simulator uses binary representation for signals. This is completely different than signal representation in an analog simulator. The digital signals can introduce instability in analog simulation due to abrupt changes in node voltage when the digital node changes the state. To take care of this problem, the interface models are inserted between the digital and analog

portions of the design. The model should consider output loading, interconnect parasitics, voltage thresholds, and interface delay.

- **Mapping states**: The digital simulators handle logic levels 0, 1, X (don't care), and Z (high impedance) states. The analog simulator requires precise voltages and currents at all nodes in a circuit. This leads to a problem of how to map the X and Z states in analog simulation. The interface models must take care of the X and Z states by representing appropriate voltages at the nodes.
- **Language**: Check for the language support provided in the simulation environment. The same language can be used for both analog and digital design descriptions and modeling. The AMS extensions of VHDL and Verilog are available as VHDL-AMS, IEEE standard 1076.1-1999, and Verilog-A/MS, standard 1.3. Some solution providers have announced that they will be releasing tools based on these language standards. Spectre®HDL, HDL-A, and Diablo are the languages defined and released by various companies for describing AMS designs.
- **Results display**: Mixed-signal design consists of both analog and digital elements. The environment should provide user-friendly result analysis capabilities, including analog and digital waveform display, and the ability to point out errors quickly.
- **User references**: References from current users of the environment is very important for understanding the problems and the features desired.
- **Cost**: Cost depends on the configuration options selected for a particular application.

4.3.2 Limitations of the Current Environments

The industry has been slow in responding to the mixed-signal design and simulation requirements. The environments currently available do not easily fit into the established chip design methodology, making it very costly to define and prove the new design methodology and effectively embed the design framework. In addition, current solutions demand high compute resources, further increasing the investment.

4.4 Using SPICE

Analog design engineers have been using the simulation program with integrated circuit emphasis (SPICE) or SPICE-like tools for analog simulation for over 30 years. SPICE is a general purpose analog simulator and contains models of circuit elements, such as resistors, capacitors, inductors, and semiconductor devices. SPICE handles complex, nonlinear circuits, and it can perform the following functions for analog designs:

- Calculate DC operating points
- Perform transient analyses
- Locate poles and zeros of transfer functions
- Perform signal transfer functions
- Determine the signal frequency response
- Determine the signal sensitivities
- Perform Fourier, distortion, and noise analysis

SPICE provides very high simulation accuracy, but the speed and design capacity handled is limited and not suitable for large designs.

Event-based simulators are suitable for digital logic, but they do not provide the required accuracy for analog simulation.

Some of the mixed-signal simulation solutions available in the industry are based on SPICE, interfaced with event-based digital simulators. Many solution providers offer their own version of SPICE tools, with features such as front-end design entry, schematic capture, comprehensive device libraries, links to digital simulators, data analysis, and graphical user interfaces (GUI).

4.5 Simulation Methodology

Figure 4-6 shows the mixed-signal simulation methodology flow that is offered in many environments. The methodology steps are as follows.

- **Enter the design**: Initially, the design is described in behavioral level or through schematic entry. The behavioral description can be in a hardware description language (AHDL) (for example, Verilog-A, SpectreHDL, or VHDL-A) for the analog part of the design, and Verilog or VHDL for the digital part. The behavioral models can be used to speed up the simulation, since the circuit-level simulation is time-consuming.

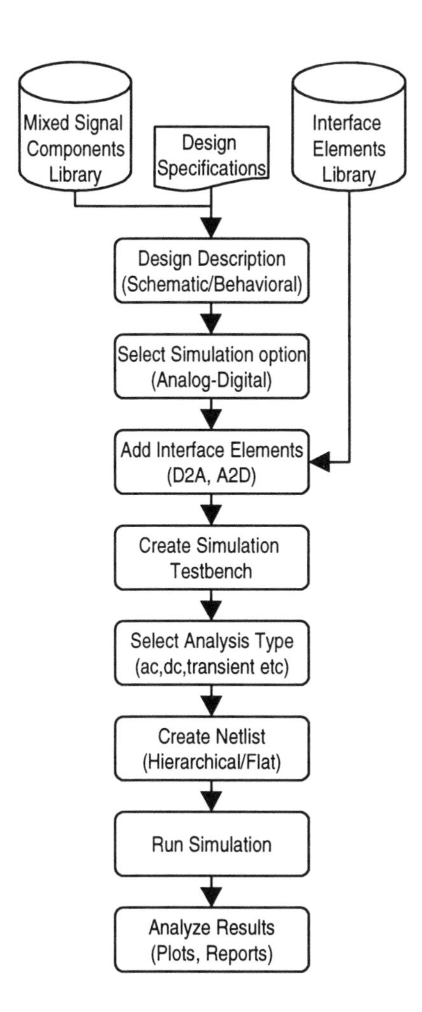

Figure 4-6. Mixed-Signal Simulation Methodology Flow

- **Select simulator options**: The simulators used in the environment provide a number of options and should be set-up for the required configuration.
- **Add interface elements**: The interface elements are associated with the digital component input or output pins. They translate signals between the analog and digital domains. A digital input pin implies translation from the analog to digital domain. A digital output pin implies translation from the digital to analog domain.

- **Create simulation testbench**: Create the testbench as per the mixed-signal design requirements. Analog test inputs are applied through voltage sources; digital test inputs are created in Verilog.
- **Select analysis type**: The appropriate analysis type for the design must be selected. Most of the mixed-signal environments are incorporated with the following analysis types: transient, AC, DC, noise, periodic AC, periodic noise, and periodic transfer functions. For example, transient analysis is suitable for mixed-signal designs.
- **Create netlist**: In some of the environments available in the industry, the netlist can be automatically created and viewed before simulation.
- **Run the simulation**: The simulation environment is executed on the design using the testbench created and selecting the appropriate simulation option.
- **Analyze the results**: The output waveforms and report files are examined to determine whether the design meets the intended functionality. The environments provide various plotting options, such as auto-plot after simulation, overlay plots, or direct plots.

4.6 Bluetooth SOC Digital-to-Analog Converter

The Bluetooth SOC uses a 16-bit DAC to reproduce the voice signal received from the Codec block. It accepts 16-bit two's complement digital input, and it outputs a signed analog voltage. It has both behavioral (Verilog-A) and schematic views. During simulation run, either of the views can be used. The mixed-signal simulation is performed using the Cadence Analog Design Environment.

Figure 4-7 shows the schematic view of the DAC. The circuit consists of a flip-flop for each digital input. The output of the flip-flop is connected through a resistor to the negative input of the amplifier. The analog output (Yout) is proportional to the digital input. Example 4-1 gives the Verilog-A code for the DAC block.

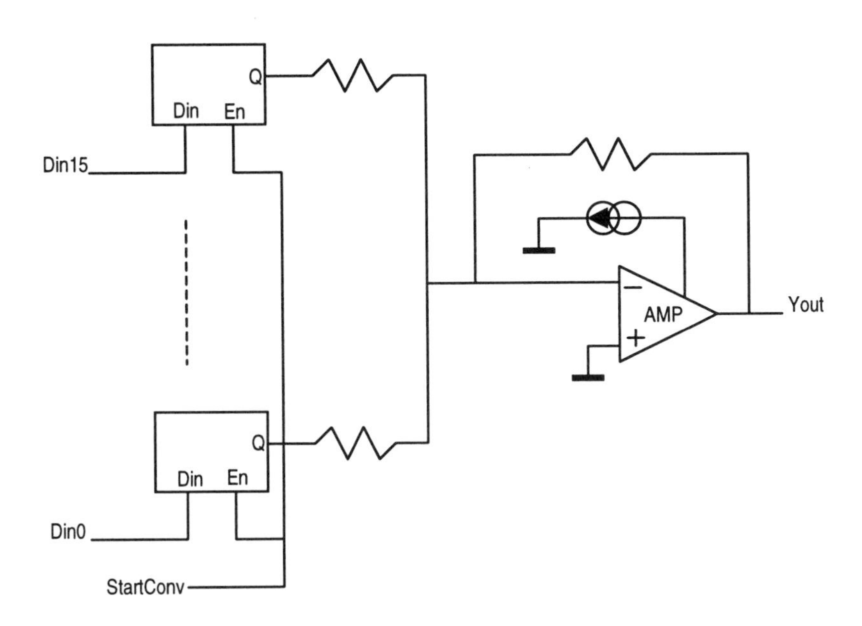

Figure 4-7. Schematic View of DAC

Example 4-1. DAC Block Description in Verilog-A

```
`include "constants.h"
`include "discipline.h"
`define VREF   5.0
module D2A(clk, Din15, Din14, Din13, Din12, Din11,
Din10, Din9, Din8,Din7, Din6, Din5, Din4, Din3, Din2,
Din1, Din0, vout);
input clk, Din15, Din14, Din13, Din12, Din11, Din10,
Din9, Din8, Din7, Din6, Din5, Din4, Din3, Din2, Din1,
Din0;
output vout;
electrical clk, Din15, Din14, Din13, Din12, Din11,
Din10, Din9, Din8, Din7, Din6, Din5, Din4, Din3, Din2,
Din1, Din0, vout;
integer x, one_detected;
integer Din[15:0];
```

```
analog begin
@(cross(V(clk)-'VREF/2.0, +1)) begin
    Din[15] = (V(Din15) > 0.0) ? 1 : 0;
    Din[14] = (V(Din14) > 0.0) ? 1 : 0;
    Din[13] = (V(Din13) > 0.0) ? 1 : 0;
    Din[12] = (V(Din12) > 0.0) ? 1 : 0;
    Din[11] = (V(Din11) > 0.0) ? 1 : 0;
    Din[10] = (V(Din10) > 0.0) ? 1 : 0;
    Din[9]  = (V(Din9)  > 0.0) ? 1 : 0;
    Din[8]  = (V(Din8)  > 0.0) ? 1 : 0;
    Din[7]  = (V(Din7)  > 0.0) ? 1 : 0;
    Din[6]  = (V(Din6)  > 0.0) ? 1 : 0;
    Din[5]  = (V(Din5)  > 0.0) ? 1 : 0;
    Din[4]  = (V(Din4)  > 0.0) ? 1 : 0;
    Din[3]  = (V(Din3)  > 0.0) ? 1 : 0;
    Din[2]  = (V(Din2)  > 0.0) ? 1 : 0;
    Din[1]  = (V(Din1)  > 0.0) ? 1 : 0;
    Din[0]  = (V(Din0)  > 0.0) ? 1 : 0;
// Convert the 2's complement input to integer and per
// form arithmetic
    if (Din[15] == 1) // sign of 2's complement i/p
    begin
     one_detected = 0;
     if (Din[0] == 1)
     begin
      one_detected = 1;
      x = 1;
     end // if (Din[0] == 1)
     else
     x = 0;
        if ((one_detected == 0 && Din[1] == 1) ||
(one_detected == 1 && Din[1] == 0))
     begin
       one_detected = 1;
       x = x + pow(2, 1);
     end // if ((one_detected == 0 && Din[1] == 1) ...
        if ((one_detected == 0 && Din[2] == 1) ||
(one_detected == 1 && Din[2] == 0))
     begin
       one_detected = 1;
       x = x + pow(2, 2);
```

```
      end // if ((one_detected == 0 && D[in2] == 1) ...
          if ((one_detected == 0 && Din[3] == 1) ||
(one_detected == 1 && Din[3] == 0)
      begin
        one_detected = 1;
        x = x + pow(2, 3);
      end // if ((one_detected == 0 && Din[3] == 1) ..
          if ((one_detected == 0 && Din[4] == 1) ||
(one_detected == 1 && Din[4] == 0))
      begin
        one_detected = 1;
        x = x + pow(2, 4);
      end // if ((one_detected == 0 && Din[4] == 1) ...
          if ((one_detected == 0 && Din[5] == 1) ||
(one_detected == 1 && Din[5] == 0))
      begin
        one_detected = 1;
        x = x + pow(2, 5);
      end // if ((one_detected == 0 && Din[5] == 1) ...
          if ((one_detected == 0 && Din[6] == 1) ||
(one_detected == 1 && Din[6] == 0))
      begin
        one_detected = 1;
        x = x + pow(2, 6);
     end // if ((one_detected == 0 && Din[6] == 1) ...
          if ((one_detected == 0 && Din[7] == 1) ||
(one_detected == 1 && Din[7] == 0))
      begin
        one_detected = 1;
        x = x + pow(2, 7);
      end // if ((one_detected == 0 && Din[7] == 1) ..
          if ((one_detected == 0 && Din[8] == 1) ||
(one_detected == 1 && Din[8] == 0))
      begin
        one_detected = 1;
        x = x + pow(2, 8);
      end // if ((one_detected == 0 && Din[8] == 1) ...
          if ((one_detected == 0 && Din[9] == 1) ||
(one_detected == 1 && Din[9] == 0))
      begin
        one_detected = 1;
```

```
        x = x + pow(2, 9);
      end // if ((one_detected == 0 && Din[9] == 1) ...
         if  ((one_detected  ==  0  &&  Din[10]  ==  1)  ||
(one_detected == 1 && Din[10] == 0))
      begin
        one_detected = 1;
        x = x + pow(2, 10);
      end // if ((one_detected == 0 && Din[10] == 1) ...
         if  ((one_detected  ==  0  &&  Din[11]  ==  1)  ||
(one_detected == 1 && Din[11] == 0))
      begin
        one_detected = 1;
        x = x + pow(2, 11);
      end // if ((one_detected == 0 && Din[11] == 1) ..
         if  ((one_detected  ==  0  &&  Din[12]  ==  1)  ||
(one_detected == 1 && Din[12] == 0))
      begin
        one_detected = 1;
        x = x + pow(2, 12);
      end // if ((one_detected == 0 && Din[12] == 1) ...
         if  ((one_detected  ==  0  &&  Din[13]  ==  1)  ||
(one_detected == 1 && Din[13] == 0))
      begin
        one_detected = 1;
        x = x + pow(2, 13);
      end // if ((one_detected == 0 && Din[13] == 1) ...
         if  ((one_detected  ==  0  &&  Din[14]  ==  1)  ||
(one_detected == 1 && Din[14] == 0))
      begin
        one_detected = 1;
        x = x + pow(2, 14);
      end // if ((one_detected == 0 && Din[14] == 1) ...
//if x is still zero, the i/p corresponds to the
//least -ve number
//else, x is only the -ve of what is already computed
      x = (x == 0) ? -pow(2, 15) : -x;
    end // if (Din[15] == 1)
    else // if (Din[15] == 0)  i/p is a +ve number
    begin
      x = (Din[14] == 1) ? pow(2, 14) : 0.0;
      x = (Din[13] == 1) ? (x + pow(2, 13)) : x;
```

```
        x = (Din[12] == 1) ? (x + pow(2, 12)) : x;
        x = (Din[11] == 1) ? (x + pow(2, 11)) : x;
        x = (Din[10] == 1) ? (x + pow(2, 10)) : x;
        x = (Din[9]  == 1) ? (x + pow(2, 9))  : x;
        x = (Din[8]  == 1) ? (x + pow(2, 8))  : x;
        x = (Din[7]  == 1) ? (x + pow(2, 7))  : x;
        x = (Din[6]  == 1) ? (x + pow(2, 6))  : x;
        x = (Din[5]  == 1) ? (x + pow(2, 5))  : x;
        x = (Din[4]  == 1) ? (x + pow(2, 4))  : x;
        x = (Din[3]  == 1) ? (x + pow(2, 3))  : x;
        x = (Din[2]  == 1) ? (x + pow(2, 2))  : x;
        x = (Din[1]  == 1) ? (x + pow(2, 1))  : x;
        x = (Din[0]  == 1) ? (x + pow(2, 0))  : x;
      end // if (Din[15] == 0)
   end // @(cross(V(clk)-'VREF/2.0, +1))
   V(vout) <+ x * 'VREF/(5.0*pow(2, 14));
end // analog
endmodule
```

4.6.1 Testbench for the DAC

Input to the DAC comes from a testbench file containing 16-bit two's complement, representing a 5KHz sine wave. It is routed through a file interface module defined in Verilog, as shown in Figure 4-8. Input can either be selected from the file or directly from the input pins. Example 4-2 shows the testbench for the DAC.

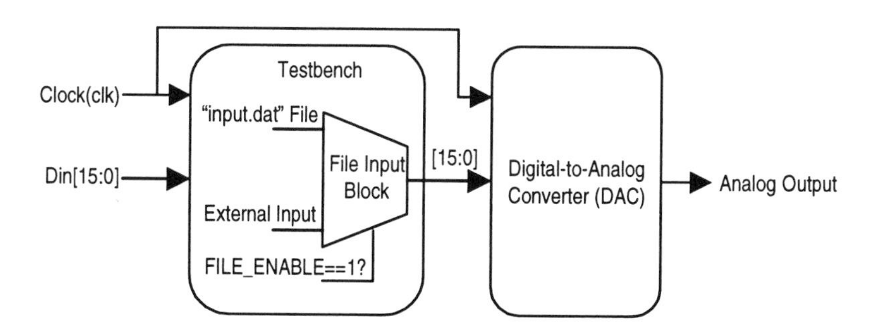

Figure 4-8. Testbench for the DAC

Example 4-2. Testbench for the DAC

```
fileInput block(Digital)  is modeled in Verilog as
follows:

`define INPUT_FILE_ENABLE  1
// Enable (1) or Disable (0) input from a file
module fileInput(clk, Din15, Din14, Din13, Din12, Din11,
Din10, Din9, Din8, Din7, Din6, Din5, Din4, Din3, Din2,
Din1, Din0, Dout15, Dout14, Dout13, Dout12, Dout11,
Dout10, Dout9, Dout8, Dout7, Dout6, Dout5, Dout4, Dout3,
Dout2, Dout1, Dout0);
input clk, Din15, Din14, Din13, Din12, Din11, Din10,
Din9, Din8,  Din7, Din6, Din5, Din4, Din3, Din2, Din1,
Din0;
output Dout15, Dout14, Dout13, Dout12, Dout11, Dout10,
Dout9, Dout8, Dout7, Dout6, Dout5, Dout4, Dout3, Dout2,
Dout1, Dout0;
reg  [15:0] mem[0:20000000];
reg Dout15, Dout14, Dout13, Dout12, Dout11, Dout10,
Dout9, Dout8, Dout7, Dout6, Dout5, Dout4, Dout3, Dout2,
Dout1, Dout0;
reg [15:0] tmp;
integer i;
initial
begin
  i = 0;
  if (`INPUT_FILE_ENABLE == 1)
   $readmemb ( "./tx_input.dat", mem ) ;
end // initial
 always @(negedge clk)
begin
  if (`INPUT_FILE_ENABLE == 1)
  begin
    tmp[15:0] = mem[i];
    Dout15 = tmp[15];
    Dout14 = tmp[14];
    Dout13 = tmp[13];
    Dout12 = tmp[12];
    Dout11 = tmp[11];
    Dout10 = tmp[10];
```

```
      Dout9   = tmp[9];
      Dout8   = tmp[8];
      Dout7   = tmp[7];
      Dout6   = tmp[6];
      Dout5   = tmp[5];
      Dout4   = tmp[4];
      Dout3   = tmp[3];
      Dout2   = tmp[2];
      Dout1   = tmp[1];
      Dout0   = tmp[0];
      i = i+1;
    end // if ('INPUT_FILE_ENABLE == 1)
    else
    begin
      Dout15 = Din15;
      Dout14 = Din14;
      Dout13 = Din13;
      Dout12 = Din12;
      Dout11 = Din11;
      Dout10 = Din10;
      Dout9   = Din9;
      Dout8   = Din8;
      Dout7   = Din7;
      Dout6   = Din6;
      Dout5   = Din5;
      Dout4   = Din4;
      Dout3   = Din3;
      Dout2   = Din2;
      Dout1   = Din1;
      Dout0   = Din0;
    end // if !('INPUT_FILE_ENABLE == 1)
  end // always @(negedge clk)
  endmodule
```

When the parameter INPUT_FILE_ENABLE is enabled (set to 1), the inputs are accessed from the tx_input.dat file. If the parameter is disabled (set to 0), the data available at the module's input terminals is used.

4.6.2 Creating the Netlist

A netlist of the design is created before starting simulation. In the SpectreVerilog simulation environment, the netlist can be created automatically and is comprised of both analog and digital netlists.

Example 4-3 shows the netlist for the D2A schematic in the SigmaDelta library. Notice that the ahdl_include directive takes the name of a file containing the behavioral description of D2A. There are instance statements for one analog-to-digital and 16 digital-to-analog interfaces (inputs to D2A).

Example 4-3. Excerpts of the Spectre Analog Netlist

```
// Design library name: SigmaDelta
// Design cell name: D2A_test
// Design view name: config

simulator lang=spectre
global 0

// BEGIN Test Fixture Interface Elements

// Analog to digital interface element

_ie99999 (clk 0) a2d dest="99999" timex=1m vl=1.5 vh=3.5

// Library name: SigmaDelta
// Cell name: D2A_test
// View name: schematic

D2A0 (clk net290 net291 net292 net293 net294 net295
net296 net297 net298 net299 net300 net301 net302 net303
net304 net305 audio_out) D2A

// BEGIN Hierarchical Interface Elements

// 16 Digital to analog interface elements at the input

_ie99983 (net302 0) d2a src="99983" fall=2n rise=3n
val1=5 val0=0
_ie99984 (net303 0) d2a src="99984" fall=2n rise=3n
```

```
val1=5 val0=0
:
:

:
:_ie99998 (net295 0) d2a src="99998" fall=2n rise=3n
val1=5 val0=0


// END Hierarchical Interface Elements

simulatorOptions options reltol=1e-3 vabstol=1e-6
iabstol=1e-12 temp=27 tnom=27 scalem=1.0 scale=1.0
gmin=1e-12 rforce=1 maxnotes=5 maxwarns=5 digits=5
cols=80 pivrel=1e-3 ckptclock=1800 sensfile="../psf/
sens.output"

//Transient analysis setup statement

tran tran stop=100u write="spectre.ic"
writefinal="spectre.fc" annotate=status maxiters=5
:
:

:
:

//Statement to include SpctreHDL behavior
ahdl_include "/SigmaDelta/D2A/veriloga/veriloga.va"
```

Example 4-4 shows the digital netlist of the DAC block.

Example 4-4. Digital Netlist

```
'timescale 1ns / 1ns
module test;
wire audio_out;
reg clk;
reg [15:0] Din;
```

```
integer dc_mode_flag;
integer output_change_count;
integer max_dc_iter;
integer dc_iterations;
time vmx_time_offset;

D2A_test top(audio_out, Din, clk);
`define verimix
`ifdef verimix
  //Parasitic Simulation annotate definitions
  `include "annotate_msb"
  //vms and dc iteration loop definitions
  `include "IE.verimix"
    //please enter any additional stimulus in the
testfixture.verimix file
  `include "testfixture.verimix"
  //$save_waveform definitions
  `include "saveDefs"
`endif

`ifdef verilog
 //please enter any additional verilog stimulus in the
testfixture.verilog file
  `include "testfixture.verilog"
`endif

`ifdef veritime
  // please enter any veritime stimulus in the
testfixture.veritime file
  `include "testfixture.veritime"
`endif

`ifdef verifault
  // please enter any verifault stimulus in the
testfixture.verifault file
  `include "testfixture.verifault"
`endif

endmodule
```

```
module D2A_test ( audio_out, Din, clk );
output  audio_out;
input  clk;
input [15:0]  Din;
specify
    specparam CDS_LIBNAME  = "SigmaDelta";
    specparam CDS_CELLNAME = "D2A_test";
    specparam CDS_VIEWNAME = "schematic";
endspecify

fileInput I0 ( clk, Din[15], Din[14], Din[13], Din[12],
Din[11],  Din[10],  Din[9],  Din[8],  Din[7],  Din[6],
Din[5], Din[4], Din[3], Din[2], Din[1], Din[0], net290,
net291, net292, net293, net294,  net295, net296, net297,
net298, net299, net300, net301, net302, net303, net304,
net305);
```

4.6.3 Simulation

Mixed-signal simulation is run on the DAC using the SpectreVerilog simulator, which performs the analog simulation with the Spectre analog simulator, and the digital simulation with the Cadence Verilog®-XL simulator.

The top-level schematic of the DAC consists of two sub-blocks: fileInterface, a digital block, and DAC, an analog block. Both the blocks are defined in behavioral models, with the digital block in Verilog, and the analog block in Verilog-A.

The DAC block has a schematic view that can be used for simulation. The schematic can be created using existing components from the AMS library supported by the tool. Once the schematic or behavioral view of the block is created, the symbol or cell view is created and used in the top-level schematic.

After the appropriate tool setup and design selection, the simulation is run using the testbench created in Verilog.

4.6.4 Response

Figure 4-9 shows the results at the output node of the DAC block. The testbench is created by digitizing a sine wave of frequency 5KHz by sampling at regular intervals. The peak-to-peak voltage of the input is four volts. The digital code generated by the testbench is applied at the input of the DAC. The output of the DAC for a half cycle (100usec) is shown in Figure 4-9.

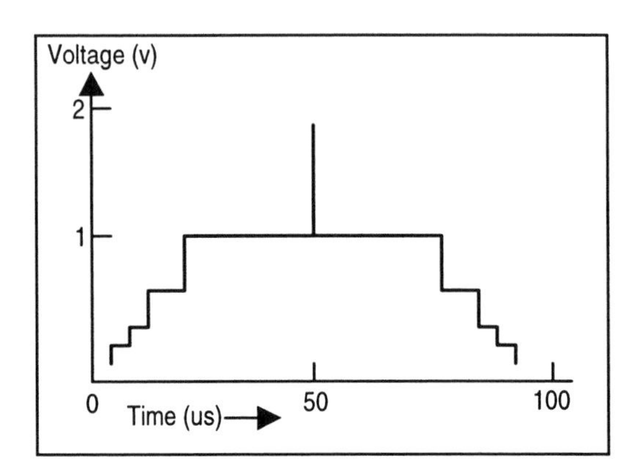

Figure 4-9. Output Response of the DAC

4.7 Chip-Level Verification with an AMS Block

Chip-level verification is performed after the AMS block is integrated. Before integration, the AMS block is verified in standalone mode.

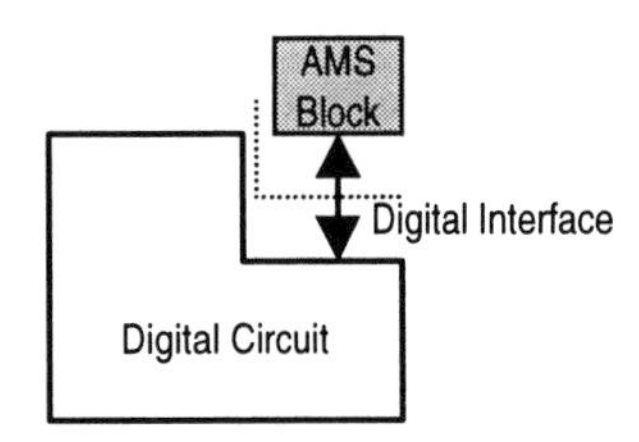

Figure 4-10. Chip-Level Verification with AMS Block

Chip-level verification can be performed in the following ways:

- **Behavioral level verification**: Using the behavioral models of the digital and AMS blocks.

- **AMS as hard block**: Verification is performed at the interface level by masking the analog pins. The interface-level simulation is digital only. The simulation checks the expected response at the digital pins.

Summary

Simulation is a particular challenge when using AMS blocks in an SOC design, and the industry has been slow in developing mixed-signal verification tools. Simulation of AMS blocks and integration with SOC can be verified using the techniques described.

References

1. Bassak Gil. Focus report: Analog and mixed-signal simulators, Integrated System Design, January, 1999.

2. Davies Phil. Design and simulation of a mixed-signal motor control system on a chip, Computer Design, August, 1999.

3. Small Charles H. Mixed-signal HDLs add analog expertise to digital simulations, Computer Design, June, 1998.

4. Specks Will J, Broderick Peter, ... A mixed digital-analog 16b microcontroller ..., IEEE International Solid-state Circuits Conference, 2000.

5. Patterson Andrew. A/MS design with a single-kernel, open environment, Electronics Engineer, August 1999.

6. Ohr Stephan. The future of mixed-signal design, Electronics Engineer, February 2000.

7. Liu Edward, Sivakumar S. Integrating analog and digital circuitry in SOC designs, Electronics Engineer, August 1998.

8. User manuals of Cadence Analog Design Environment.

CHAPTER 5

Simulation

In large chip designs, functional verification or simulation often takes nearly 40 to 70 percent of the overall effort of a dedicated team of verification engineers. Functional simulation has become a major bottleneck in product development, affecting both the cost and time-to-market constraints. Smart simulation techniques are required to overcome these problems.

This chapter addresses the following topics:

- Functional simulation
- Testbench wrappers
- Event-based and cycle-based simulations
- Simulating an ASB/APB Bridge
- Transaction-based verification
- Simulation acceleration

The simulation concepts are illustrated with the Bluetooth SOC design example.

5.1 Functional Simulation

Complex system-on-chips (SOC) built using pre-verified, reusable intellectual property (IP) blocks pose new simulation challenges. It takes a lot of resources to create testbenches to verify each IP block exhaustively. Integrating blocks that are already verified can introduce additional integration problems. To verify the block integration, the design should be verified using the following tests:

- **Block-to-block interconnect verification**: To perform this, a set of tests that exercises the complete functionality of the IP blocks in a system must be developed. The tests should be self-checking to allow automatic checking of the response, which eases the regression testing efforts required later. The simulation should also focus on the bus interconnect, between individual IP blocks and memory and registers read/write, and transactions between the IP blocks.
- **Bus contention test**: The design should be verified for contention by various blocks for data transfer on the same bus. There may be situations where multiple masters contend for the bus simultaneously. This requires testing the arbiter block and direct memory access (DMA) blocks to ensure the correct functionality.
- **Interface protocol/compliance test**: Many SoCs use industry bus standards, such as PCI, IEEE-1394, IEEE-1284, inter IC bus (I2C), serial port interface (SPI), Smart card, and ARM-AMBA to interface the system. This requires that protocol or compliance testing be performed and checking with the IP provider whether the IP meets the intended compliance tests.

As the amount of verification required for complex designs continues to increase, it is necessary to create verification environments and improve simulation performance to achieve time-to-market objectives. The simulation performance will improve as event-based simulation give way to cycle-based simulation and acceleration techniques, such as hardware accelerators, emulation, and rapid prototyping systems.

Figure 5-1 shows a methodology flow for functional simulation. The methodology assumes the system register-transfer level (RTL) code as the input. The RTL code goes through lint checking for syntax and synthesizability checks. Formal model checking is performed to verify behavioral properties of the blocks in the design. The model checking tool uses constraints and properties of the design as inputs.

RTL functional simulation uses the system testbench. Event-based simulation, cycle-based simulation, transaction-based verification, or simulation acceleration can be run on the design depending on the simulation requirements.

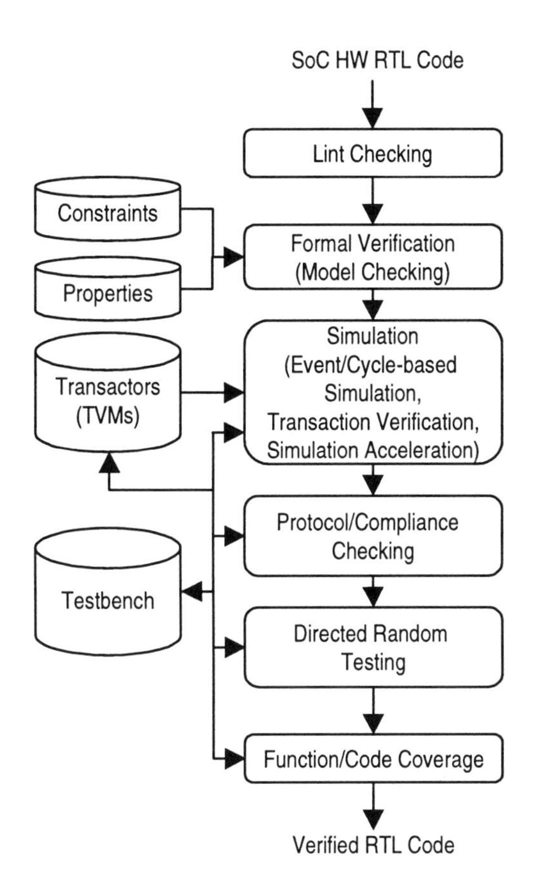

Figure 5-1. SOC Hardware Simulation

Protocol/compliance testing identifies any protocol violations that appear on the bus. Directed random testing checks the corner cases in the control logic. These use probabilistic distributing functions, Poisson and uniform, which enable simulation using real-life statistical models. Code coverage identifies any untested paths in the design.

This chapter illustrates functional simulation. Lint checking, model checking, protocol checking, directed random testing, and code coverage are described in Chapter 3, "Block-Level Verification."

5.2 Testbench Wrappers

The first step in a simulation is to create an environment that reflects the architecture of the target design. This task occurs during the system design phase of the project. The simulation environment is interface-centric and is created by wrapping each block or component to be integrated into the design with a testbench wrapper. This provides a variety of capabilities, which are described in this section.

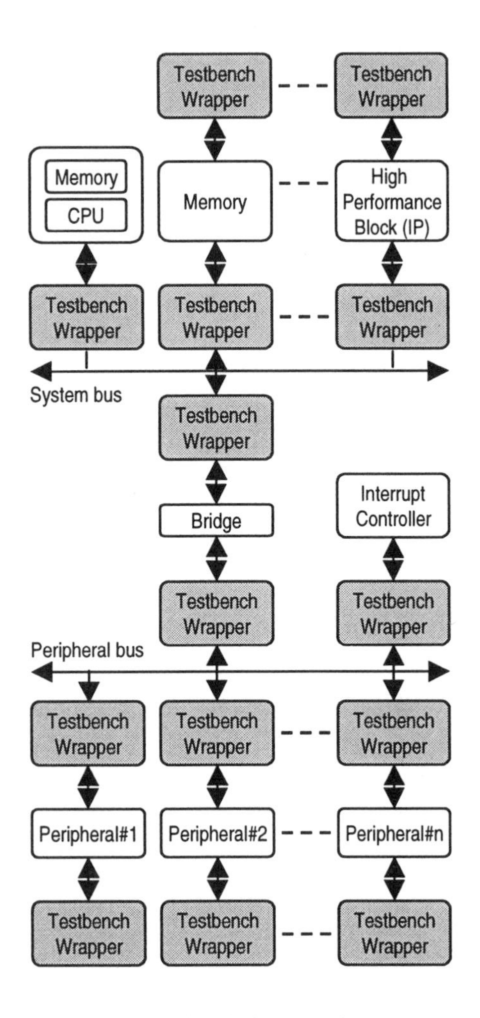

Figure 5-2. Simulation Environment

Figure 5-2 illustrates the simulation environment for a design. Testbench wrappers encapsulate each IP block in an SOC design. By wrapping each block, a wrapper for the complete system can also be created. A block-level wrapper can be used to perform standalone verification, which provides more control over the identification of bugs or errors. This testbench wrapper is not part of the design implementation but exists purely to facilitate design verification. Testbench wrappers can perform the following functions:

- Drive test suites into their associated block or system
- Drive test suites out onto the associated buses
- Operate in a transparent mode where patterns generated from other wrappers pass through the block wrapper unaltered

Testbench wrappers use vector suites from the following:

- Previously generated block- or system-level tests, such as those supplied by the IP provider
- Directed random patterns, which are well suited for bus protocol checking
- Manually created test suites
- Vector suites captured during the execution of a system-level test

Figure 5-3 shows a simple block diagram of a testbench wrapper. Testbench wrappers provide the following additional functionality.

- **Test vector translation**: Transforms vector sets from one format to another, enabling the same vector suite source to exercise different models of a design. For example, a frame-based video image can be transformed into a pin-accurate, serial-bit stream.
- **Random pattern generation**: Used to do the following:

  -Tests the control logic

  -Enables protocol checkers to validate bus transactions

  -Use probabilistic distribution functions, Poisson and uniform, which enable simulation using real-life statistical models
- **Code coverage**: Assesses the functional coverage achieved by a particular testbench when applied to a design. This can be at the individual block level or the full-chip level.

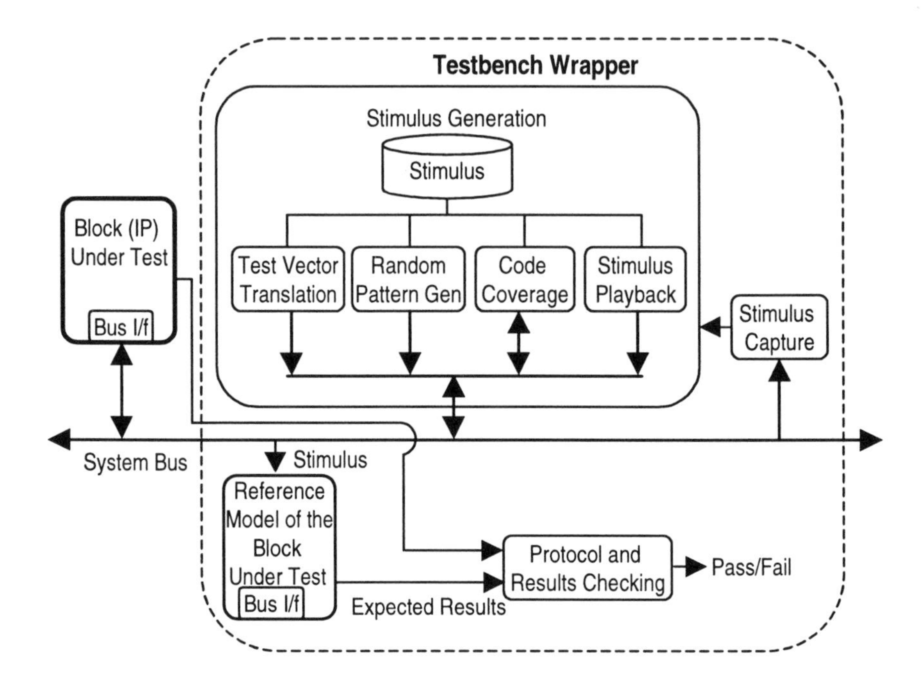

Figure 5-3. Block with a Testbench Wrapper

- **Stimulus capture**: Captures a vector set for the testbench wrapper's associated block during a test driven from an external source. For example, a block-level test can be captured when executing a system-level test, allowing the block to be simulated in isolation at a later time.
- **Stimulus playback**: Plays back testbench vectors captured during simulation on the block to test the functionality of the block in isolation.
- **Results checking**: Checks results in one of the following two ways:

  -Compares the results of a simulation to a previously specified, expected response file

  -Exercises two different views of a design simultaneously and checks that the responses are equivalent
- **Protocol/compliance checking**: Checks for any bus protocol violations that appear on the testbench wrapper's bus interfaces during simulation.

5.2.1 Block Details of the Bluetooth SOC

Figure 5-4 shows a simple block diagram of the example Bluetooth SOC design (details of the design are explained in Chapter 1).

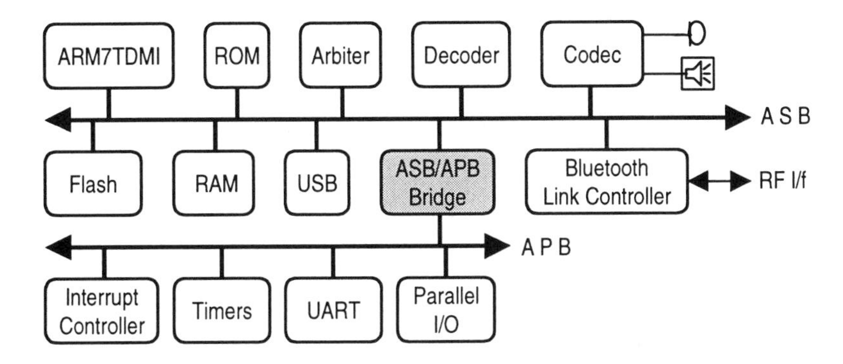

Figure 5-4. Block Diagram of the Bluetooth SOC

The testbench wrapper is illustrated with the advanced system bus (ASB) slave example. The ASB-to-advanced peripheral bus (ASB/APB) bridge block is used to illustrate event-based simulation and cycle-based simulation.

5.2.2 Test Vector Translation

Test vector translation involves converting a vector from a high-level format to pin-accurate and cycle-accurate bits that can be applied to the inputs of a device at detailed design level. The translation depends on the input format of the vector and the format of the desired final output.

A common translation technique involves loading data into memory and applying the memory data, byte-by-byte or word-by-word to the input pins. When serial data is applied, a parallel byte stream or word stream is left-shifted or right-shifted accordingly.

Example 5-1 shows how to convert a data stream stored in bytes to serial stream data that is applied to a pin on each rising edge of the clock.

Example 5-1. Test Vector Translation

```
integer i;
wire [7:0] loadData;
// Data from loadData is loaded to regData
reg  [7:0] regData;   // stores 8 bit data
reg  INPUT_PIN;// serial input data
wire DataEn;
// DataEn is a pin which signals that data can be
//applied to the device.
reg  [3:0] count;// A counter to count
always @(posedge CLK)
 begin
   if (! bnres)
    count = 4'b0000;// initializing the counter
   else
    if (DataEn && (count != 4'b1000))
     begin
       INPUT_PIN = loadData[0];

// applying the serial data
       loadData = loadData >> 1;
// shifting the contents of the register
       count = count + 1;
      end
     else
       begin
         count = 0;
         loadData = regData;
       end  // DataEn
 end  // always begin
```

5.2.3 Stimulus Generation

Stimulus generators drive stimulus directly onto the bus, either by previously created vectors from the database or by randomly generated vectors. The appropriate combination of stimulus generators would allow any block or sub-block of the design to be stimulated for verification.

The previously created vectors in the database cover testing of all possible features and conditions. The vectors are written based on the specification (for example, the data sheet of IP block). All boundary conditions are exercised. These vectors are created to check that the model covers exception cases in addition to the regular functional cases. Test vectors are broken down into smaller sets of vectors, each testing a feature or a set of features of the model. These test vectors are built incrementally. For example:

- Test vectors initially developed for basic functions, for example, read/writes, reset, and decode tests
- Various data patterns, for example, all ones and zeroes 0xff, 0x00, walking ones and zeroes 0xaa, 0x55
- Deterministic boundary test vectors, for example, FIFO full and empty tests
- Test vectors of asynchronous interactions, including clock/data margining
- Bus conflicts, for example, bus arbitration tests in a full SOC tests
- Implicit test vectors that may not be mentioned in the data sheet, for example, action taken by a master on a bus retract
- System boot
- System multitasking and exception handling
- Randomly generated full-system tests

The following routines generate bus cycles on the ASB in the Bluetooth SOC design. They emulate the external memory access and the address and data bus interface using their processor controls. Examples of the `write_burst` and `read_burst` routines are given below.

- Non-sequential write (`write_arm`)
- Non-sequential read (`read_arm`)
- Sequential (burst) write (`write_burst`)
- Sequential (burst) read (`read_burst`)

5.2.3.1 write_burst Routine

The `write_burst` routine writes to successive locations in memory. Inputs are an address, the number of locations to be written, and data. If the lowest bit of the data is "z," a random number is generated and written to the memory location, and a copy is maintained in the local images of the memories. Otherwise, the data supplied to this routine is used.

Example 5-2. `write_burst` Routine

```
Task/Function Name: write_burst
Synopsis: write_burst (addr, num, data)
addr:  Address of memory location for a write burst
operation.num : The number of locations to write into.
data: The data that has to be written into this location
task write_burst;

input [31:0] addr;
input        num;
input        data;
reg   [31:0] data;
integer      num;
integer       i;

begin
  'ifdef arm_model
     'ifdef thumb
// execute instructions for memory transfers in the
// thumb mode
     'else
// execute instructions for memory transfers in the
//regular mode
     'endif
  'else
     nMREQ  = 1'b0;
     SEQ    = 1'b0;
     nRW    = 1'b0;
     nWAIT  = 1'b1;
     APE    = 1'b0;
     DBE    = 1'b1;
     for (i = 0; i < num; i = i + 1)
      begin
       if (i == 0)
        begin
          @ (negedge MCLK)
            fork
       {nMREQ, SEQ} = #('Tmsh) 2'b01;
        nWAIT       = #('Tmclkp/2.0 - Tws) 1'b0;
        nRW         = #('Tape) 1'b1;
```

```
        A              = #('Tape) addr;
        DBE            = 1'b0;
        join
      end
     @ (negedge MCLK)
       fork
        APE           = #('Taph) 1'b1;
        nWAIT         = #('Twh) 1'b1;
        DBE           = 1'b1;
        D_OUT         =  #('Tdout) data;
        A             = #('Tmclkp - 'Tape) (addr + 4);
        join
     local_memory[addr] = data;
// store a local copy of the data in a memory element
   end
 'endif
```

5.2.3.2 read_burst Routine

The `read_burst` routine reads from successive locations in memory. It uses the address and number of locations as its inputs. The data on the bus is compared against the data that is stored locally in the images of the memory.

Example 5-3. `read_burst` Routine

```
Task/Function Name: read_burst
Synopsis: read (addr, num)
addr: Address of memory location for a read burst opera-
tion.
num  : The number of data locations to read from.

task read_burst_arm;
input  [31:0]    addr;input num;
reg    [31:0]       data;

begin

  'ifdef arm_model
     'ifdef thumb
```

```
// execute instructions for memory transfers in the
//thumb mode
      `else
// execute instructions for memory transfers in the
// regular mode
      `endif
  `else
       nMREQ    = 1'b0;
       SEQ      = 1'b0;
       nRW      = 1'b0;
       nWAIT    = 1'b1;
       APE      = 1'b1;
       DBE      = 1'b1;
       A        =  addr;
       @ (negedge MCLK)
         fork
             {nMREQ, SEQ}  = #(`Tmsh) 2'b01;
             nWAIT         = #(`Tmclkp/2.0 - Tws) 1'b0;
             DBE           = 1'b0;
         join
       @ (negedge MCLK)
         fork
             APE           = #(`Tmclkp - `Taps) 1'b0;
             nWAIT         = #(`Twh) 1'b1;
             DBE           = 1'b1;
         join
       for (i = 0; i < num; i = i + 1)
           begin
              @ (negedge MCLK)
                data = local_memory[addr];
// retrieve the local copy of the memory
                check_data_arm (data);
                if (i != 0)
                    A  =  (addr + 4);
          end
  `endif
end
endtask //end read_burst_arm
```

5.2.4 Stimulus Capture

The verification team is responsible for determining the set of test vectors or test patterns that adequately test the functionality of a design. These test patterns are an explicit indication of the sequence of transitions of every external signal of the design. While generating the system-level testbench at a high level (C or behavioral), the boundary patterns or vectors can be captured for the IP blocks. These boundary patterns can be handed over to the IP development team so that the IP is tested for integration to the system while it is being developed. In this way the system-level stimulus gets propagated down to the block-level testbenches using the boundary-capture approach.

To generate test vectors for RTL code for application to the post-layout netlist or custom circuit, the example routine written in Verilog can be used.

Example 5-4. Verilog Routine for Stimulus Capture

```
input  agnt, bclk, bnres, DA, bwait;
output areq, CONTN, BA, bd, LOUT;
module test_biu;
reg    areq, CONTN,  BA, bd, LOUT;
wire   agnt, bclk, bnres, DA, bwait;
//top level core instance
initial
  begin
       // stimulus
  end
parameter INPUTS  = {agnt, bclk, bnres, DA, bwait };
parameter OUTPUTS = {areq, CONTN, BA, bd, LOUT};
// Creating input and output files
integer infile, outfile;
initial
begin
  infile   = fopen("inputVectors");
  outfile = fopen("outputVectors");
end

// If you want to monitor these signals just after a
positive clock edge
always @(posedge clk) begin
#1;
$fdisplay(infile, "%b%b%b%b", INPUTS);
```

```
// Or
//$fdisplay(infile, "%b%b%b%b%b",agnt, bclk, bnres,
//DA,bwait);
$fdisplay(outfile, "%b%b%b%b%b", OUTPUTS);
//$fdisplay(outfile, "%b%b%b%b%b", areq, CONTN, BA, bd,
//LOUT);
end
endmodule
```

These vector files can be updated based on any change in input or output, in which case the `always` block changes to the following:

```
always @(agnt, bclk, bnres, DA, bwait)
begin
  $fdisplay(infile, "%b%b%b%b%b", INPUTS);
end
always @(areq, CONTN, BA, bd, LOUT)
begin
  $fdisplay(outfile, "%b%b%b%b%b", OUTPUTS);
end
```

Based on the application, capturing vectors can be automated by writing scripts to generate these Verilog files for the set of signal transitions to be captured. Knowledge of the model's functionality is needed for identifying the set of test vectors required.

5.2.5 Results Checking

Results checking is a commonly used approach to leverage a reference model and then compare the device under verification (DUV) to the reference model.The reference model is typically implemented either in C or a hardware description language (HDL) behavioral model. C language provides a powerful feature to verify the algorithm of the model and validate the system intent. The C model can be used as a reference model for RTL and structural design verification. The C testbench used for verifying the C model can be reused in the HDL testbench.

Example 5-5 shows the `check_data` routine for an ARM processor.

Example 5-5. `check_data` Routine

```
Task/Function Name: check_data_***
Synopsis           : check_data_arm (compare_data)
Description :
task check_data_arm;input [31:0] compare_data;
begin
 `ifdef arm_model
   `ifdef thumb
// execute instructions for memory transfers in the
//  thumb mode
   `else
// execute instructions for memory transfers in the
//  regular mode
   `endif
 `else
   if (D_IN !== compare_data)
    begin
    $display("Data violation at %d ns", $time);
     $display("At address %h, compare data is %h and %h
was read " , A, compare_data, D_IN);
    end
end
  `endif
endtask //end check_data_arm
```

5.2.6 Testbench Wrapper for a Slave

This section illustrates a testbench wrapper for a slave using the Bluetooth SOC design. The ASB slave model consists of the ASB slave and the addressing locations, including ROM, RAM, and FIFO.

A slave wrapper is application-specific, depending on whether the slave is FIFO, ROM, RAM, or a communication device. The wait states on the master are determined by addressing an out-of-page location in the master. Similarly, the error states are determined by writing to a FIFO-full condition. The last condition is resolved by addressing the last location of a memory page on the slave. The following examples show the bus protocol checker and wrapper for the ASB slave.

Example 5-6. Testbench Wrapper for a Slave Block

```
'include "slave_watch.h"
  module slave_wrapper;
  reg BCLK;
  reg BnRES;
  reg DSEL;
  reg ['asb_addr_width - 1 :0] BA;
  reg              BWRITE;
  reg [1:0]        BSIZE;

  // Assumption : When BSIZE represents data
  transfers less than 32 bits long,
  // only the first few bits will be used.

  reg ['asb_data_width - 1:0] SLAVEDATA_IN;
  reg WAIT_IN;
  reg ERROR_IN;
  reg LAST_IN;
  wire BWAIT;
  wire BERROR;
  wire BLAST;

  // Generate Clock circuitry
  initial
  begin
   BCLK = 0;
   BnRES = 0;
   forever
   begin
      #('period/2) BCLK = ! BCLK;
   end
  end

  // Generate reset
  initial
   fork
    begin
    #('reset_cycles * 'period) ;
     @(negedge BCLK) BnRES = 1;
    end
  // The slave is selected in this address
   #(('reset_cycles +2 ) * 'period)
```

```
 write_arm (addr, data);  // For address =
32'h8000_0000 ; data = 32'haaaa_aaaa

// The slave is not selected in this address
 #(('reset_cycles +4 ) * 'period)
 write_arm (addr, data);  // For address =
32'h8800_0000 ; data = 32'haaaa_aaaa
// The slave is not selected in this address

 #(('reset_cycles +6 ) * 'period)
  read_arm (addr, data);  // For address =
32'h8000_0000 ; data = 32'haaaa_aaaa
// The slave is not selected in this address

 #(('reset_cycles +6 ) * 'period)
  read_arm (addr, data);  // For address =
32'h8800_0000 ; data = 32'haaaa_aaaa
// Do a burst operation
#(('reset_cycles +8 ) * 'period)
write_burst (addr,10, $random);  // doing 10 burst
cycles with random data ; addr = 32'h8000_000f
#(('reset_cycles +18 ) * 'period)
read_burst (addr,10, $random);  // doing 10 burst
cycles with random data; addr = 32'h8000_000f

 // *** do bursts with addr and num taking random
values.
// ** condition the test for BWAIT , BERROR, BLAST to
be flagged.
  #('timeout_cycles * 'period) $finish;
join
endmodule
```

The following module is a verification watcher for a slave. Watching can be turned off by setting `stop_watching` to 1. All signals are referenced using hierarchical referencing.

Example 5-7. Slave Watcher

```
'include "slave_watch.h"
module slave_watch (stop_watching);
input stop_watching;
wire BCLK;
wire BnRES;
wire ['asb_addr_width -1 :0] BA;
wire ['asb_data_width -1 :0] BD;
wire BWRITE;
wire [1:0] BSIZE;
wire [1:0] BPROT;
wire DSEL;
wire BWAIT;
wire BERROR;
wire BLAST;
wire  data_tristate_en;
wire  resp_tristate_en;
wire  wait_condition;
wire  last_condition;
wire  error_condition;

// Following code must be inserted for
errors_code_incl.v to work.
integer severity;
integer errorsCount;

// Severity = 0, implies a fatal error does not meet the
spec
// Severity = 1, implies a error did not meet the
implementation requirements.
// Severity = 2, implies the primary signals do not meet
the required specs.
// Severity = 3, Unintended z or x.
wire  verbose;
wire  quiet;
reg  [7:0] continue_error;

// EOC required for cut and paste of
errors_code_incl.v
assign BCLK = 'slave_clk;
assign BnRES = 'slave_reset;
assign BA = 'slave_ba;
```

```
assign  BD = 'slave_bd;
assign  BWRITE = 'slave_bwrite;
assign  BSIZE = 'slave_bsize;
assign  BPROT = 'slave_bprot;
assign  DSEL = 'slave_dsel2;
assign  BWAIT = 'slave_wait;
assign  BERROR = 'slave_error;
assign  BLAST = 'slave_last;
assign wait_condition = 'slave_waitcond;
assign last_condition = 'slave_lastcond;
assign error_condition = 'slave_errorcond;
assign verbose = 'slave_verbose_flag;
assign quiet = 'slave_quiet_flag;

// All the response signals and Data bus must be
//tristated when the slave
// is not selected.
'ifdef check_tristate
 always @ (DSEL or BnRES or BTRAN)
 if (! stop_watching);
 begin
  if (! DSEL)
  begin
   if (! quiet)
   begin
    if (BD !== 'bz)
     $write ("ERROR: Slave cycle -- Data not
tristated@time %t DSEL=0\n",$t
ime);
    if ({BWAIT,BERROR,BLAST} !== 'bz)
     $write ("ERROR: Slave cycle -- Response signals
not tristated @time %t
 DSEl = 0 \n", $time);
   end // if quiet
  end // if DSEL
  if (! BnRES)
  begin
   if (! quiet)
   begin
    if (BD !== 'bz))
```

```
     $write ("ERROR: Slave cycle -- Data not tristated
during reset @time%t
\n", $time);
    if ({BWAIT,BERROR,BLAST} !== 'bz) )
     $write ("ERROR: Slave cycle -- Response signals
not tristated @time %t
\n", $time);
   end // if quiet
  end // if BnRES
 if (BTRAN == 2'b00)
  begin
   if (! quiet)
   begin
    if (BD !== 'bz))
     $write ("ERROR: Slave cycle -- Data not tristated
during ATRAN @time %
t\n", $time);
    if ({BWAIT,BERROR,BLAST} !== 'bz) )
     $write ("ERROR: Slave cycle -- Response signals
not tristated @time %t
\n", $time);
   end // if quiet
  end // if BnRES
 end // @DSEL
`endif
 initial
  errorsCount = 0;
always @(posedge BCLK)
 if (! stop_watching)
 begin
  # (`period/2 - `resolution) ;
  if (wait_condition )
  begin
   if (!BWAIT)
   begin
    severity = 0;
    errorsCount = errorsCount + 1;
    if (! quiet)
     $write ("ERROR: Slave cycle - BWAIT not set @time
%t\n", $time );
   end // if ! BWAIT
```

```
    else
    begin
     if ((verbose == 1) && (quiet == 0 ))
      $write ("Passed -- wait implemented properly\n");
    end // BWAIT
   end // wait_condition
   if (last_condition )
   begin
    if (!BLAST)
    begin
     severity = 0;
     errorsCount = errorsCount + 1;
     if (! quiet)
      $write ("ERROR: Slave cycle - BLAST not set @time
%t\n", $time );
    end // if ! BLAST
    else
   else
    begin
     if ((verbose == 1) && (quiet == 0 ))
      $write ("Passed -- last implemented properly\n");
    end // BLAST
   end // last_condition
   if (error_condition )
   begin
    if (!BERROR)
    begin
     severity = 0;
     errorsCount = errorsCount + 1;
     if (! quiet)
      $write ("ERROR: Slave cycle - BERROR not set @time
%t\n", $time );
    end // if ! BERROR
    else
    begin
     if ((verbose == 1) && (quiet == 0 ))
      $write ("Passed -- error implemented
properly\n");
    end // BERROR
   end // error_condition
 end // always
```

```
// This code exits the simulation based on the error
level
 always @(errorsCount)
  case (severity)
   0 : if (continue_error >= 0 )
     begin
      $write ("level 0 errors
encountered...exiting\n");
      $finish;
     end
   1 : if (continue_error == 1 )
     begin
      $write (" level 1 errors
encountered...exiting\n");
      $finish;
     end
   2 : if (continue_error == 2 )
     begin
      $write (" level 2 errors
encountered...exiting\n");
      $finish;
     end
   default : severity = 255;
  endcase

// This code exits the simulation based on the error
level
 always @(errorsCount)
  case (severity)
   0 : if (continue_error >= 0 )
     begin
      $write ("level 0 errors
encountered...exiting\n");
      $finish;
     end
   1 : if (continue_error == 1 )
     begin
      $write (" level 1 errors
encountered...exiting\n");
      $finish;
```

```
        end
      2 : if (continue_error == 2 )
        begin
         $write (" level 2 errors
encountered...exiting\n");
         $finish;
        end
      default : severity = 255;
     endcase
endmodule
```

The following file must be included in the slave watcher. It contains information on the signals and conditions required by the watcher.

Example 5-8. Slave Watcher Conditions

```
'define asb_addr_width 32
'define asb_data_width 32
'define number_of_slaves 4
'define period 10
'define resolution 1
'define timeout_cycles 10000
'define reset_cycles 5
'define slave_test slave_test
'define slave_clk 'slave_test.CLK
'define slave_reset 'slave_test.BnRES
'define slave_ba 'slave_test.BA
'define slave_bd 'slave_test.BD
'define slave_bwrite 'slave_test.BWRITE
'define slave_wait 'slave_test.BWAIT
'define slave_last 'slave_test.BLAST
'define slave_error 'slave_test.BERROR
'define slave_bsize 'slave_test.BSIZE
'define slave_bprot 'slave_test.BPROT
'define slave_dsel2 'slave_test.DSEL2
'define slave_waitcond (BA == 32'h8000_0fff) //
exceeds page boundary on master
```

```
'define slave_lastcond (BA == 32'h8000_7ffc) //
exceeds th last location on the slave.
'define slave_errorcond ((BA >= 32'h8000_0000) && (BA
< 32'h8000_0100) )&& BWRITE
'define data_en_type !
'define resp_en_type !
'define slave_verbose_flag 0
'define slave_quiet_flag 0
```

5.3 Event-based Simulation

Event-based simulation simulates digital circuits (designs) by taking events, one at a time, through the design until a steady state condition is achieved. As shown in Figure 5-5, Input_1 and Input_2 are fed to the digital logic circuit under test, which produces Output. The EBS tool computes the steady state Output for every event on Input_1 and Input_2.

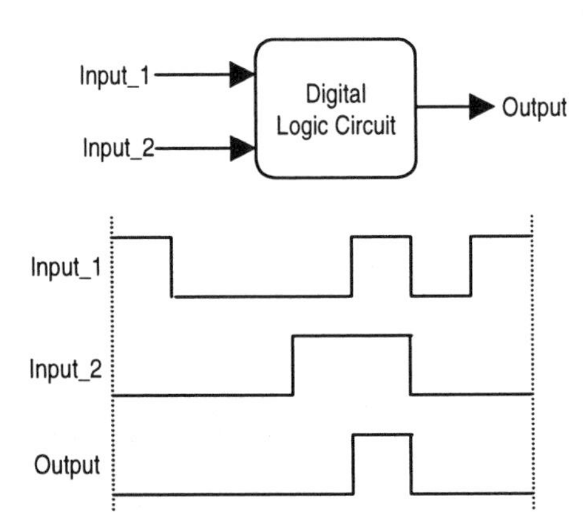

Figure 5-5. Event-based Simulation

In EBS, the design models include timing and functionality. Any change in input stimulus is identified as an event and propagated through each stage in the design. A design element may be evaluated several times in a single cycle due to the different arrival times of the inputs and the feedback of signals from downstream design elements. While this provides a highly accurate simulation environment, the speed of the execution depends on the size of the design, and it can be relatively slow for large designs.

EBS tools can support 4 to 28 states and the simulation of designs described in the following representations:

- Behavior of design described in HDL
- RTL code
- Gate level
- Transistor level

5.3.1 Types of EBS Tools

The two types of EBS tools that are available in the industry are:

- **Compiled-code simulator**: Accepts the design described in HDL, compiles the design to a data structure, and runs it like any other executable program on the host machine. Examples are Cadence® NC-Verilog® and Verilog Compiled Simulation (VCS) simulator.
- **Interpreted-code simulator**: Accepts the design described in HDL, interprets each line of code, and runs it on the host machine. An example is Cadence Verilog-XL.

Compiled-code simulators are faster than interpreted-code simulators.

5.3.2 EBS Environment

The EBS environment consists of the following main elements, as shown in Figure 5-6.

- Facility to describe the design. The design can be described in RTL. The environment also accepts gate-level netlists as input.
- Input design code parser to check the code.
- Testbench created by the user.
- Event-based simulator.
- Facility to analyze the results obtained after simulation. The results could be in the form of reports and waveforms.

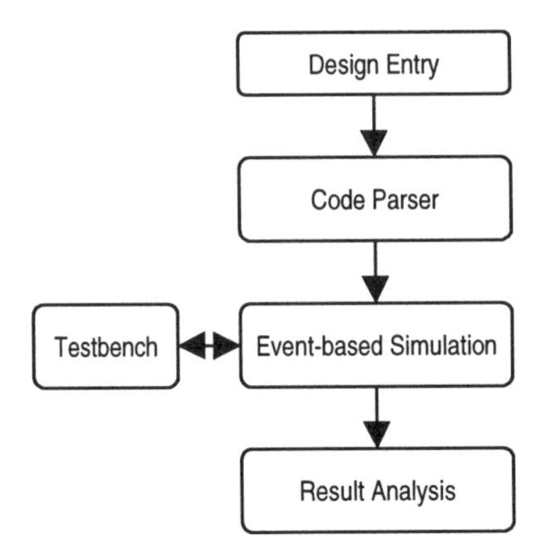

Figure 5-6. Event-based Simulation Environment

5.3.3 Selecting an EBS Solution

Some of the issues to consider when selecting an EBS solution are:

- **Capacity**: What design complexity the tool can handle, the memory required for the host machine, and compile time.
- **Performance**: Most users find that performance is the major bottleneck in project schedules. Performance depends upon design size, compile time, link time, execution time, memory utilization, and the number of test vectors.
- **Compliance check**: Should handle all (or most of) the language compliance aspects that are defined by the standards. For example, for Verilog the standard is Open Verilog International's (OVI) Verilog Language Reference Manual (Verilog LRM), and the programming language interface (PLI).
- **Debugging capability**: Which debugging features are provided. Debugging helps in identifying the errors in the design quickly. The simulation output can be in the form of waveforms or report files.
- **Design environment interface**: Should provide easy interfaces with hardware accelerators, emulators, modelers, libraries, and so on.
- **Support**: Check for complete documentation, technical support, online access for downloading and updating programs, and training.

5.3.4 EBS Methodology

The EBS methodology flow, as shown in Figure 5-7, is as follows:

1. **Design acceptance**: The code checker determines whether the design is acceptable to the EBS tool.
2. **Testbench creation**: Create the testbench as per the design specifications, using Verilog, VHDL, C with PLI, or any other suitable verification language, such as Vera or Specman Elite.
3. **Simulation**: Run the event-based simulator using the RTL design and the testbench.
4. **Debugging**: The output is analyzed and errors fixed, after which the simulation is rerun.

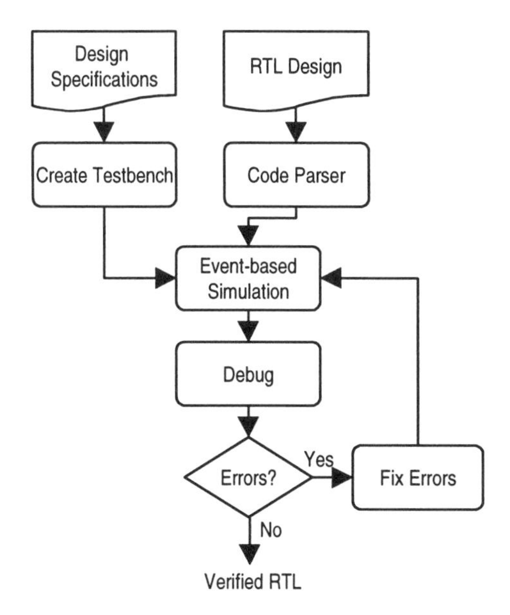

Figure 5-7. Event-based Simulation Methodology Flow

5.4 Cycle-based Simulation

Cycle-based simulation (CBS) simulates the digital circuits by computing the steady-state response of the circuit at the end of each clock cycle, as shown in Figure 5-8. The clocked circuits are also called as synchronous circuits.

CBS does not consider the timing aspects of the design, so a separate static-timing analysis tool for verifying the timing is required. CBS tools are typically faster than EBS tools since the output response is computed at the clock edges instead of computing the output for a change in any of the inputs as in EBS. Some CBS tools handle only 0 and 1 logic states and do not consider the X and Z states in the digital circuits. This further increases the simulation speed and enables faster design iterations.

CBS tools use considerably less memory in the host machine compared to event-based simulators, so larger circuits can be simulated.

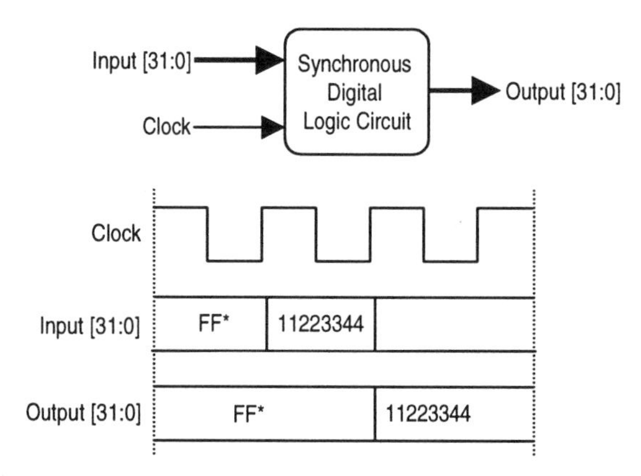

Figure 5-8. Cycle-based Simulation

5.4.1 When to Use CBS

CBS is a big boon for simulating very large designs (>100K ASIC gates) that could easily take months to simulate using the fastest event-based simulators available. CBS is also highly suitable for extensive regression testing. Some of the applications where CBS can be used are:

- Microprocessor design
- Complex floating-point processing block designs.
- Symmetric multiprocessing design.
- Digital signal processing (DSP) chips
- SoCs with processors and peripherals
- Large designs with multiple clock domains
- Printed circuit board/module designs.

5.4.2 CBS Environment

The CBS environment consists of the following main elements, as shown in Figure 5-9.

- Facility to describe the design. The design can be described in RTL. The environment also accepts gate-level netlists as input.
- Input design code style checking facility to check design code acceptable to CBS.
- Testbench created by the user.
- Cycle-based simulator.
- Facility to analyze the results obtained after simulation. The results could be in the form of reports and waveforms.

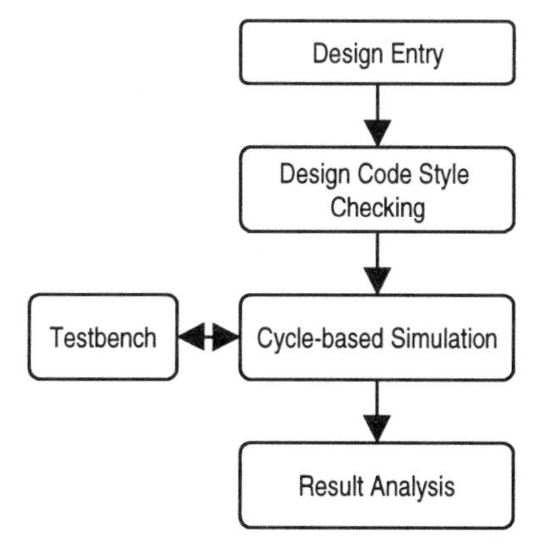

Figure 5-9. Cycle-based Simulation Environment

5.4.3 Selecting a CBS Solution

Some of the issues to consider when selecting a CBS solution are:

- **Capacity**: What design complexity the tool can handle, the memory required for the host machine, and compile time.
- **Performance**: Solution providers claim that performance is 10x to 100x times that of EBS for a given synchronous design.
- **Multiple clocks**: Should be able to simulate designs with multiple and independent clock domains.
- **Asynchronous design**: Must be able to handle asynchronous logic and performance, since most designs contain a small portion of asynchronous logic.
- **X and Z states**: Logic circuits can have 0,1, X, and Z states. Most CBS tools only take care of 0 and 1 states in order to achieve better simulation speed, but the X and Z states are required for knowing the state of the logic during initialization. Otherwise, it may be necessary to perform EBS during initialization of the design.
- **Existing design**: It is preferable to use the CBS with few or no modifications in the existing design.

5.4.4 Limitations of CBS

CBS has the following limitations:

- For asynchronous designs, CBS performance is poor.
- CBS does not take timing of the design into consideration, hence timing verification needs to be performed using a static-timing analysis tool.
- Most of the commercially available CBS tools places restrictions on the coding style and language constructs used to describe the design.
- Some of the CBS tools do not handle X and Z states in logic. This requires EBS to be used for initialization.
- CBS performance depends upon the level of activity within the design. For designs and test suites with low levels of activity (events) EBS can provide greater performance.

5.4.5 CBS Methodology

CBS methodology accepts RTL design or gate-level netlist as input. Figure 5-10 shows the CBS methodology flow.

- **Design acceptance**: Checks whether the RTL code is suitable for CBS. This may be performed by a tool available with a CBS solution. The code style checker examines whether the design is acceptable for CBS.
- **Testbench creation**: A testbench not tuned for CBS can degrade the overall simulation performance. The ideal testbench for CBS is a synthesizable RTL and should be synchronous. The testbench can be created in Verilog, VHDL, C with PLI, or any other suitable verification language. Some of the tools automatically generate the testbench required for CBS by accepting testbenches that are written for EBS.
- **Simulation**: Runs cycle-based simulator using the RTL design and testbench.
- **Debugging**: Analyze the output after simulation and fix error, after which the simulation is rerun.

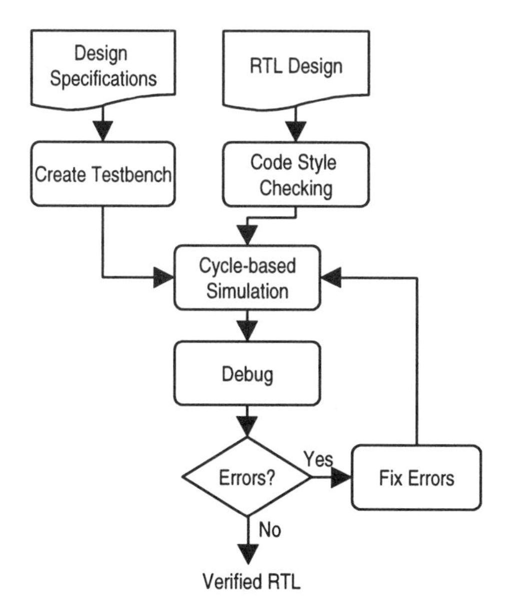

Figure 5-10. Cycle-based Simulation Methodology Flow

5.4.6 Comparing EBS and CBS

Table 5-1 shows the comparison between EBS and CBS technology.

Table 5-1. Comparison between EBS and CBS

Parameter	EBS	CBS
Output evaluation is done	For every input change	For every clock cycle
Computation of time delay	Performed	No
Event scheduling is done	Yes	No
Store timing information	Yes	No
Identify timing violations	Yes	No
Simulation speed	Low	10 to 100x times EBS
Static timing analysis tool required for timing analysis	Not required	Required
Best application scenario	Simulation of blocks and critical part of a large design	Highly suitable for design regression testing

5.5 Simulating the ASB/APB Bridge

This section illustrates EBS and CBS using the ASB/APB bridge block in the Bluetooth SOC. The simulation uses the Cadence NC-Verilog tool for EBS and the SpeedSim™ tool for CBS.

5.5.1 ASB/APB Block

The ASB/APB bridge is a slave on an ASB and translates ASB transfers into a suitable format for the slave devices on the APB. It latches the address, data, and control signals, and performs address decoding to generate slave select signals for peripherals on the APB. Figure 5-11 shows the signals connected to the bridge.

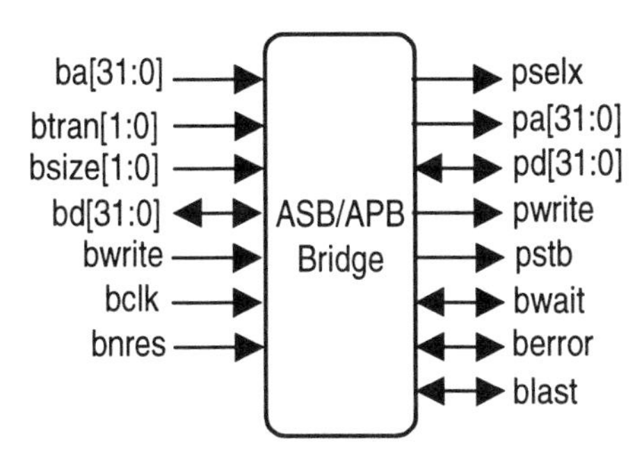

Figure 5-11. Block Diagram of ASB/APB Bridge

The following signals are connected to the ASB/APB bridge:

- ba—ASB address bus
- btran—ASB transfer type indication signals
- bsize—ASB transfer size (8 bits, 16 bits, or 32bits)
- bd—ASB data bus
- bwrite—ASB read/write control signal (0 means read; 1 means write)
- bclk— ASB bus clock signal
- bnres—ASB reset signal (active low)
- pselx—APB peripheral select signals
- pa— APB address bus
- pd—APB data bus
- pwrite—APB read/write control signal (0 means read; 1 means write)
- pstb—APB peripheral strobe signal
- bwait—ASB wait signal
- berror—ASB error signal
- blast—ASB last transfer signal

5.5.2 Design RTL Code

Example 5-9 gives the RTL code for the ASB/APB bridge in Verilog. The clock is defined in a way that is acceptable by both EBS and CBS tools.

Example 5-9. RTL Code for ASB/APB Bridge in Verilog

```
module asb_apb_bridge
       ( BA, BD, BERROR, BLAST, BSIZE, BWAIT, BWRITE,
         BnRes, DSEL, PA, PD, PSEL, PSTB, PWRITE );
input             BWRITE;
input             BnRes;
input             DSEL;
input    [1:0]    BSIZE;
input    [31:0]   BA;
inout    [31:0]   BD;
inout    [31:0]   PD;

output            BERROR;
output            BLAST;
output            BWAIT;
output            PSTB;
output            PWRITE;
output   [31:0]   PA;
output   [31:0]   PSEL;

'ifdef SPEEDSIM
// For Cycle-based Simulation (CBS)
              wire BCLK;
              SPDclkgen2 #5 (.phi1(BCLK));
            'else
// For Event-based Simulation (EBS)
              reg BCLK;
              initial BCLK = 1;
              always begin
                #5 BCLK = ~BCLK;
              end
            'endif


wire     [1:0]    BSIZE;
wire     [31:0]   BA;
wire     [31:0]   BD;
wire     [31:0]   PD;
wire     [31:0]   PSEL_r;
wire     [31:0]   PA_r     = (DSEL == 1'b1)? BA : 32'hz;
```

```
wire    [31:0]  PD_dsl  = (DSEL == 1'b1)? BD : 32'bz;
wire            pw_r    = (DSEL == 1'b1)? BWRITE : 1'b0;
wire    [31:0]  BD_psl = (PSEL != 32'h0000_0000)? PD :
32'hz;

reg             PWRITE, PSTB;
reg     [1:0]   pos_cnt, neg_cnt;
reg     [31:0]  PSEL, PA, BD_r, BD_r2, PD_r, BD_bsz,
PD_bsz;
reg             b_wait_r, b_last_r, b_erro_r;

// PSEL for 32 devices - decoder
assign PSEL_r[0]  = (BA[24:20] == 5'b00000)? 1'b1 :
1'b0;
assign PSEL_r[1]  = (BA[24:20] == 5'b00001)? 1'b1 :
1'b0;
assign PSEL_r[2]  = (BA[24:20] == 5'b00010)? 1'b1 :
1'b0;
assign PSEL_r[3]  = (BA[24:20] == 5'b00011)? 1'b1 :
1'b0;
assign PSEL_r[4]  = (BA[24:20] == 5'b00100)? 1'b1 :
1'b0;
assign PSEL_r[5]  = (BA[24:20] == 5'b00101)? 1'b1 :
1'b0;
assign PSEL_r[6]  = (BA[24:20] == 5'b00110)? 1'b1 :
1'b0;
assign PSEL_r[7]  = (BA[24:20] == 5'b00111)? 1'b1 :
1'b0;
assign PSEL_r[8]  = (BA[24:20] == 5'b01000)? 1'b1 :
1'b0;
assign PSEL_r[9]  = (BA[24:20] == 5'b01001)? 1'b1 :
1'b0;
assign PSEL_r[10] = (BA[24:20] == 5'b01010)? 1'b1 :
1'b0;
assign PSEL_r[11] = (BA[24:20] == 5'b01011)? 1'b1 :
1'b0;
assign PSEL_r[12] = (BA[24:20] == 5'b01100)? 1'b1 :
1'b0;
assign PSEL_r[13] = (BA[24:20] == 5'b01101)? 1'b1 :
1'b0;
assign PSEL_r[14] = (BA[24:20] == 5'b01110)? 1'b1 :
```

```
1'b0;
assign PSEL_r[15] = (BA[24:20] == 5'b01111)? 1'b1 :
1'b0;
assign PSEL_r[16] = (BA[24:20] == 5'b10000)? 1'b1 :
1'b0;
assign PSEL_r[17] = (BA[24:20] == 5'b10001)? 1'b1 :
1'b0;
assign PSEL_r[18] = (BA[24:20] == 5'b10010)? 1'b1 :
1'b0;
assign PSEL_r[19] = (BA[24:20] == 5'b10011)? 1'b1 :
1'b0;
assign PSEL_r[20] = (BA[24:20] == 5'b10100)? 1'b1 :
1'b0;
assign PSEL_r[21] = (BA[24:20] == 5'b10101)? 1'b1 :
1'b0;
assign PSEL_r[22] = (BA[24:20] == 5'b10110)? 1'b1 :
1'b0;
assign PSEL_r[23] = (BA[24:20] == 5'b10111)? 1'b1 :
1'b0;
assign PSEL_r[24] = (BA[24:20] == 5'b11000)? 1'b1 :
1'b0;
assign PSEL_r[25] = (BA[24:20] == 5'b11001)? 1'b1 :
1'b0;
assign PSEL_r[26] = (BA[24:20] == 5'b11010)? 1'b1 :
1'b0;
assign PSEL_r[27] = (BA[24:20] == 5'b11011)? 1'b1 :
1'b0;
assign PSEL_r[28] = (BA[24:20] == 5'b11100)? 1'b1 :
1'b0;
assign PSEL_r[29] = (BA[24:20] == 5'b11101)? 1'b1 :
1'b0;
assign PSEL_r[30] = (BA[24:20] == 5'b11110)? 1'b1 :
1'b0;
assign PSEL_r[31] = (BA[24:20] == 5'b11111)? 1'b1 :
1'b0;

assign BWAIT    = (~BCLK && DSEL && BnRes) ? b_wait_r :
1'bz;
assign BLAST    = (~BCLK && DSEL && BnRes) ? b_last_r :
1'bz;
assign BERROR   = (~BCLK && DSEL && BnRes) ? b_erro_r :
```

```
1'bz;
assign BD        = (BCLK)? BD_r2 : 32'hz;
assign PD        = PD_r;

// Assignment from ASB data bus to APB data bus depend
ing upon bus size
always @(BSIZE or PD_dsl) begin
        case (BSIZE)
        2'b00 : begin
                PD_bsz[7:0]      <= PD_dsl[7:0];
                PD_bsz[15:8]     <= PD_dsl[7:0];
                PD_bsz[23:16]    <= PD_dsl[7:0];
                PD_bsz[31:24]    <= PD_dsl[7:0];
        end
        2'b01 : begin
                PD_bsz[15:0]     <= PD_dsl[15:0];
                PD_bsz[31:16]    <= PD_dsl[15:0];
        end
        2'b10 : begin
                PD_bsz           <= PD_dsl;
        end
        2'b11 : begin
                PD_bsz           <= 32'hz;
        end
        endcase
end

// Assignment from APB data bus to ASB data bus depend
ing upon bus size
always @(BSIZE or BD_psl) begin
        case (BSIZE)
        2'b00 : begin
                BD_bsz[7:0]      <= BD_psl[7:0];
                BD_bsz[15:8]     <= BD_psl[7:0];
                BD_bsz[23:16]    <= BD_psl[7:0];
                BD_bsz[31:24]    <= BD_psl[7:0];
        end
        2'b01 : begin
                BD_bsz[15:0]     <= BD_psl[15:0];
                BD_bsz[31:16]    <= BD_psl[15:0];
        end
```

```
        2'b10 : begin
                BD_bsz          <= BD_psl;
        end
        2'b11 : begin
                BD_bsz          <= 32'hz;
        end
        endcase
end

// Assignment at posedge of clk for APB signals PA, PD,
PSELx, PWRITE
always @(posedge BCLK) begin
        if (BnRes == 1'b0)      begin
                BD_r2   <= 32'hz;
                PA      <= 32'hz;
                PD_r    <= 32'hz;
                PSEL    <= 32'h0000_0000;
                PWRITE  <= 1'b0;
                pos_cnt    <= 2'b00;
                end
        else    begin
          case (pos_cnt)
          2'b00 : begin
                if (DSEL == 0)
                     begin
                        pos_cnt <= 2'b00;
                        PSEL    <= 32'h0000_0000;
                        PA      <= 32'hz;
                        PWRITE  <= 1'b0;
                        PD_r    <= 32'hz;
                        BD_r2   <= 32'hz;
                     end

                else begin
                        pos_cnt <= 2'b01;
                        PSEL    <= PSEL_r;
                        PA      <= PA_r;
                        PWRITE  <= pw_r;
                        BD_r2   <= 32'hz;
                       if (pw_r == 0) PD_r      <= 32'hz;
                       else           PD_r     <= PD_bsz;
```

```
                    end
                end

        2'b01 : begin
                        pos_cnt <= 2'b10;
                        PSEL    <= PSEL_r;
                        PA      <= PA_r;
                        PWRITE  <= pw_r;
                        PD_r    <= PD_bsz;
                        BD_r2   <= 32'hz;
                end

        2'b10 : begin
                        pos_cnt <= 2'b11;
                        PSEL    <= 32'h0000_0000;
                        PA      <= 32'hz;
                        PWRITE  <= 1'b0;
                        PD_r    <= 32'hz;
                        if  (pw_r == 0) BD_r2   <= BD_r;
                       else             BD_r2   <= 32'hz;
                end

        2'b11 : begin
                        pos_cnt <= 2'b00;
                        PSEL    <= 32'h0000_0000;
                        PA      <= 32'hz;
                        PWRITE  <= 1'b0;
                        PD_r    <= 32'hz;
                        BD_r2   <= 32'hz;
                end
        endcase
      end
end

// Assignment at negedge of clock for APB Signal PSTB
always @(negedge BCLK) begin
      if (BnRes == 1'b0)      begin
            BD_r    <= 32'hz;
            PSTB    <= 1'b0;
            b_erro_r<= 1'b0;
            b_last_r<= 1'b0;
```

```
            b_wait_r<= 1'b0;
            neg_cnt <= 2'b00;
    end
    else                        begin
      case (neg_cnt)
      2'b00 : begin
            if (DSEL == 0) begin
                    neg_cnt <= 2'b00;
                    BD_r    <= 32'hz;
                    PSTB    <= 1'b0;
                 end
            else begin
                    neg_cnt <= 2'b01;
                    BD_r    <= 32'hz;
                    PSTB    <= 1'b0;
                 end
              b_erro_r<= 1'b0;
              b_last_r<= 1'b0;
              b_wait_r<= 1'b1;
             end

      2'b01 : begin
              neg_cnt <= 2'b10;
              BD_r    <= 32'hz;
              PSTB    <= 1'b1;
              b_erro_r<= 1'b0;
              b_last_r<= 1'b0;
              b_wait_r<= 1'b1;
              end

      2'b10 : begin
              neg_cnt <= 2'b11;
              BD_r    <= BD_bsz;
              PSTB    <= 1'b0;
              b_erro_r<= 1'b0;
              b_last_r<= 1'b0;
              b_wait_r<= 1'b0;
      end

      2'b11 : begin
              neg_cnt <= 2'b00;
```

```
                    BD_r     <= 32'hz;
                    PSTB     <= 1'b0;
                    b_erro_r<= 1'b0;
                    b_last_r<= 1'b0;
                    b_wait_r<= 1'b0;
                    end
            endcase
        end
end
endmodule
```

5.5.3 Testbench for EBS

The testbench for the ASB/APB bridge is shown in Example 5-10.

Example 5-10. Testbench for ASB/APB Bridge

```
module test;
reg BnRes;
reg [31:0] BA;
wire [31:0] BD,PA,PD;
asb_apb_bridge APB_bridge( BA, BD, BERROR, BLAST, BSIZE,
BWAIT, BWRITE, BnRes, DSEL, PA, PD, PSEL, PSTB, PWRITE
);
initial
begin
 BnRes = 0;
 #20 BnRes = 1;
 #10 BA = 32'hF340F000;
 #5;
 $display("PSEL_R = %h", APB_bridge.PSEL_r);
 #5 $stop;
end
endmodule
```

5.5.4 Running a Simulation

The simulation is run on the ASB/APB bridge RTL design using Cadence NC-Verilog for EBS and SpeedSim for CBS.

Running the Cadence NC-Verilog tool on the ASB/APB bridge RTL design using a testbench gives the following results:

```
Design hierarchy summary:
                     Instances  Unique
Modules:                 2        2
Registers:              17       17
Scalar wires:           41        -
Vectored wires:          9        -
Always blocks:           5        5
Initial blocks:          2        2
Cont. assignments:       6       41
Pseudo assignments:      2        2
Writing initial simulation snapshot: worklib.test:v
Loading snapshot worklib.test:v .................... Done
ncsim> source /net/ccvob-sj2/vobstore/repo/cds/LDV2.2/991213/tools/inca/files/nc
simrc
ncsim> run
PSEL_R = 00100000
Simulation stopped via $stop(1) at time 40 NS + 0
```

The results can be analyzed using the waveform viewer incorporated with the simulation tool. The report files can be checked for errors.

The Cadence SpeedSim tool is used to perform CBS on the ASB/APB bridge. Since there is no concept of timing in CBS, SpeedSim ignores all "#" delays in the design. All clocks are defined with the SpeedSim primitive `SPDclkgen2`. In the example, the primary clock signal `BCLK` is described as a phase 1 clock that is high in the first phase and low in the second phase.

SpeedSim is run in two steps, as follows:

Step 1: Compile the design and generate the <design>.spd file. This is done by the speedbld script:

```
speedbld -s asb_apb_bridge.v
SpeedBld v3.4.0  06/23/2000 09:52
Compiling source file asb_apb_bridge.v
    "asb_apb_bridge" is a top-level module.
Latches:32       Flops:297      Logic:1641     Assert:0
Total:1970
Information: Found 32 state devices with non-exclusive
drivers.
Information:  Found  36  state  devices  with  exclusive
loads.
Information: Found 32 state devices with non-exclusive
loads.
    Final BDE Cells: 9
    Wrote asb_apb_bridge.spd (68664 bytes)

Compilation Complete  (1.1 real  0.3 user  0.0 sys)
```

Step2: Run the simulator using the <design>.spd file. Following is a snapshot of the window during the simulation run.

```
speedsim asb_apb_bridge.spd
SpeedSim v3.4.0  06/23/2000 09:55
Copyright 2000 Quickturn, a Cadence Company.
0> ls
    $0: asb_apb_bridge
0> cd asb_apb_bridge
0> ex B*
    $0: asb_apb_bridge.BA[31:0] = 00000000
    $1: asb_apb_bridge.BCLK = 0
    $2: asb_apb_bridge.BD[31:0] = 00000000
    $3: asb_apb_bridge.BD_bsz[31:0] = 00000000
    $4: asb_apb_bridge.BD_psl[31:0] = 00000000
    $5: asb_apb_bridge.BD_r[31:0] = 00000000
    $6: asb_apb_bridge.BD_r2[31:0] = 00000000
    $7: asb_apb_bridge.BERROR = 0
    $8: asb_apb_bridge.BLAST = 0
    $9: asb_apb_bridge.BSIZE[1:0] = 0
    $10: asb_apb_bridge.BWAIT = 0
    $11: asb_apb_bridge.BWRITE = 0
```

```
    $12: asb_apb_bridge.BnRes = 0
0> ex PS*
    $13: asb_apb_bridge.PSEL[31:0] = 00000000
    $14: asb_apb_bridge.PSEL_r[31:0] = 00000001
    $15: asb_apb_bridge.PSTB = 0
0> dep BA F340F000
0> step 5
5> ex BA
    $16: asb_apb_bridge.BA[31:0] = F340F000
5> ex PSEL_r
    $17: asb_apb_bridge.PSEL_r[31:0] = 00100000
5> exit
```

Named signals are examined using the **ex** command. The **ls** command lists the path levels. The **dep** command is used for depositing values during the simulation run. On bus signal BA, the value F340F000 is deposited using the **dep** command. As a result, PSEL_r gives the value 00100000 after running the simulation for five steps. The simulator can also read all the commands from a command file.

The results can be analyzed with the waveform viewer incorporated in the simulation tool.

5.6 Mixed-Event/Cycle-based Simulation

Many IP blocks provided by third-party vendors or through design reuse may not be synchronous. To convert designs and testbenches to run on a CBS tool may require more time than the simulation speed-up gained using EBS justifies. The ideal solution in such a situation is to use CBS where possible, and EBS for the remaining part of the design. Some of the solution providers have incorporated the features of both EBS and CBS in a single tool, enabling efficient mixed-event/cycle-based verification.

5.7 Transaction-based Verification

Overall productivity can be enhanced by performing verification at the system or transaction level. A transaction is a single transfer of data or control between a transactor and the design over an interface. It can be as simple as a memory read or as complex as the transfer of an entire structured data packet. Transaction-based verification (TBV) tools enhance productivity by allowing the user to perform verification at transaction level in addition to the signal/pin level.

The TBV environment provides the following features:

- Generates self-checking and parameterizable testbenches at system or transaction level
- Enhances Verilog and VHDL capabilities through the use of C++, which speeds test creation and maximizes reuse
- Testbenches can be authored in Verilog HDL or C++
- Records, displays, and analyzes the response at system or transaction level in addition to signal/pin level
- Debugging capability
- Provides functional coverage of the design
- Directed random testing to test the corner cases
- Ability to reuse testbenches
- Ability to partition and develop transaction verification modules (TVM) and tests by different verification team members.

Currently, very few solution providers offer TBV. This section illustrates the methodology associated with Cadence's TBV solution, using the Bluetooth SOC design.

5.7.1 Elements of TBV

The basic elements of the TBV environment are:

- **Design under test (DUT)**: An RTL or gate-level description of the design that is to be verified.
- **Transaction**: A single transfer of data or control between a transactor and the design over an interface that begins and ends at specific times. It is used to annotate waveform and can be associated with property information. Also the transaction can flag errors.

- **Transaction verification module**: A collection of tasks that executes a particular kind of transaction. Each TVM can generate high-level transactions, for example, do_write, do_read, do_burst_write, do_burst_read, expect_write, send_data. The transactor connects to the DUT at a design interface. Most designs have multiple interfaces, hence, they have multiple transactors to drive stimulus and check results. TVM is also referred to as a transactor or bus functional model (BFM). TVMs can be authored in HDL or C++. TVMs created for a design can be reused for other designs involving the same or similar design blocks.
- **Test**: A program that generates sequences of tasks for each of the transactors in the system. The test program invokes transactor tasks indirectly and triggers the DUT. Tests can be authored in HDL or C++.

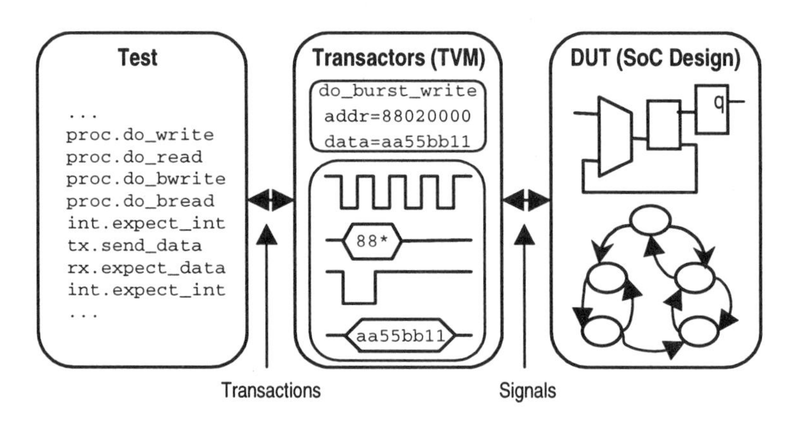

Figure 5-12. Transaction-based Verification Elements

As shown in Figure 5-12, the DUT is driven by the TVMs on its design interface. The design interface is basically a set of signals and buses. Usually a single finite state machine (FSM) controls the signals and buses in the design being verified. The sequence of states that is traversed during activity occurring on the interface consists of transactions. Examples of transactions include read, write, and packet transfers.

TVMs are driven by tests according to the system test requirements. The tests can control randomness of transaction sequence based on a previous transaction sequence, system rate, or coverage information.

5.7.2 TBV Environment

Figure 5-13 shows the TBV environment. The DUT is driven by the TVMs. Each interface requires a TVM. As explained earlier, the TVMs in turn are driven by the test (stimulus). The tests contain a set of transactions created as per the design functionality.

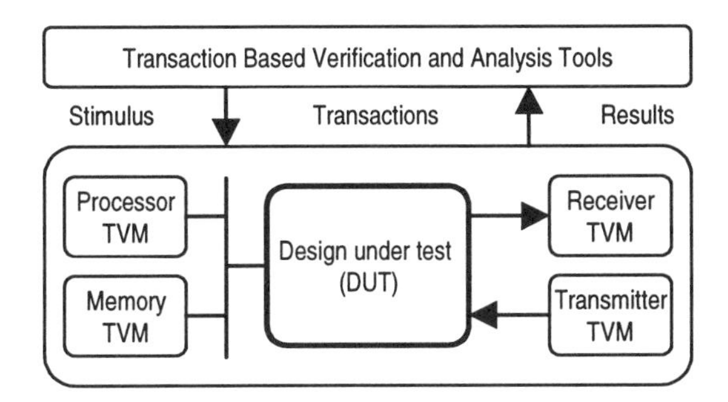

Figure 5-13. Transaction-based Verification Environment

For example, the transmitter TVM sends a data packet to the DUT. The processor TVM is interrupted by the DUT indicating the data availability. The processor TVM generates a transaction to write the received data into a known memory location. At this time, the memory TVM is set to read data from the processor TVM. At the receiver interface, a transaction can also be created to compare the actual results of data transmitted with the expected results, and results checking can be embedded within the testbench this way. These transactions are analyzed to check the DUT's functionality using the waveform analysis capability that is provided in the TBV environment. Transactions can be generated in a directed random fashion to test the DUT for corner cases.

5.7.3 Creating a Testbench

Currently, testbenches are usually written in Verilog, VHDL, or verification languages. As design complexity increases, designers need more efficient methods of generating testbenches. Many users have developed in-house C/C++ techniques to take advantage of the advanced programming capabilities of these languages. In a TBV environment, testbenches can be created in HDL or C++.

The testbench consists of a set of TVMs and tests created as per the design requirements. Example 5-11 creates the TVM and test for the ARM7TDMI processor's "non-sequential write" task.

Example 5-11. Pseudocode for the Processor's "non-sequential write"

```
begin tvm task
get address and data for the write transaction
wait for 7 unit of time //example
assert signals request and seq
wait for 5 unit of time
set signals address/mas to the appropriate address/ data
assert signal read_write
wait for 5 unit of time
set signal dreg
while (signal wait is 0)
wait for positive edge of the signal clk
wait for 2 unit of time
de-assert control signals
release data signals to high-impedance states
end
```

Example 5-12 shows the pseudocode for the test embedding the above transaction.

Example 5-12. Pseudocode for Test Incorporating the "non-sequential write" Transaction

```
begin test
select a non-sequential write transaction
with an interesting address and data.
call task do_NonSeqW
with the selected transaction.
end
```

5.7.3.1 Creating Transactions in Verilog

Transactions in Verilog can be created by calling the following transaction recording system task (TRST) calls. All these calls, except $set_property, are associated with user defined properties.

- $trans—Records simulation time and scope.
- $trans_begin and $trans_end—Define hierarchical transactions. Every $trans_begin call must have a corresponding $trans_end call.
- $trans_error—Defines error transactions.
- $set_property—Assigns properties to the current or last transaction.

Example 5-13 shows a "non-sequential write" transaction written in Verilog using the above TRST calls. It is based on the pseudocode explained in Example 5-11.

Example 5-13. "non-sequential write" Transaction in Verilog

```
task d0_ NonSeqW;
input [31:0] addr;
input [1:0] whb;
input [31:0] data;
begin
        $trans_begin("NonSeq Write cycle", "Writing
     data",
               "addr=",addr,
               "data=",data,
               "whb=",whb);
        if (whb == 2'b11)
        $trans_error("Wrong data length","Expected word
     half-word or byte",
                 "Wrong_whb=",whb);
        $trans("set_cycle_type","Initalize Cycle Type");
        #SIGNAL_D0
           nMREQ=0;
           SEQ=0;
        $set_property("nMREQ=",nMREQ,"SEQ=",SEQ);
        $trans("addr_rdy","Assert address and data
     size",
              "addr=",addr,
```

```
        "whb=",whb);
    #(SIGNAL_D - SIGNAL_D0)
        A=addr;
        nRW=1;
        MAS=whb;
$trans("data_rdy","sends data and wait for slave
nWAIT",
        "data=",data);
    #(SIGNAL_D1 - SIGNAL_D)
        D_reg=data;
    @(posedge MCLK) while (nWAIT==0) @(posedge MCLK);
    $trans("NonSeqWr_cycle_end","Terminate cycle
sequentially");
    #(SIGNAL_D0 - ASB_D)
        nMREQ=1;
        SEQ=0;
    #(SIGNAL_D - SIGNAL_D0)
        A=32'hffffffff;
        nRW=1;
        MAS=2'b11;
    #(SIGNAL_D1 - SIGNAL_D)
        D_reg=32'hzzzzzzzz;
    #(ASB_CYC - SIGNAL_D1) // line up with ASB cycle
boundary
        D_reg=32'hzzzzzzzz;
    $trans_end;
end
endtask
```

The test incorporating the above transaction is shown in Example 5-14.

Example 5-14. Test Using "non-sequential write" Transaction in Verilog

```
initial
begin
  $recordsetup("design=arm7tdmi");
  $recordvars("depth=0");
$recordfile("simple.trn", "sequence");
    ApplyInitValue;
```

```
        #(ASB_CYC - SIGNAL_D)
        #(10*ASB_CYC)
        // === The testbench start here ===
        /* setup INTC and PIT */
        NonSeqW(32'hA0200000, 2'b10, 32'hFFFFFFFF);
        NonSeqW(32'hA0100008, 2'b10, 32'hFFFFFFFF);
    //Configuring Slot control blocks for transmit at
    adress A4000000
    NonSeqW(32'hA4000000, 2'b10, 32'hF0000BE5);
        /* Introduce error by putting whb*/
        NonSeqW(32'hA0100008, 2'b11, 32'hFFFFFFFF);
        /* turn on 3-ws on Flash Controller */
    // === The testbench end here ===
        $stop;
    end
    endmodule
```

5.7.3.2 Creating Transactions in C++

Writing transactions in C++ enables easy development of transaction-based testbenches and TVMs. The C++ environment maintains the same semantic conventions as Verilog and VHDL. The C++ syntax enables the designer to create more complex tests than can be done in HDL, using dynamic processes (spawning), queues, data structures, and semaphores, but the environment behaves just like HDL.

To write a test in C++, the following is required:

- Identify the TVMs that are needed
- Make calls to the tasks in the TVMs
- Use the `$tbv_tvm_connect()` call in the HDL module to register the TVM
- Use the `$tbv_main()` call in the HDL module to start the C++ test

Example 5-15 shows the “non-sequential write” transaction in C++.

Example 5-15. "non-sequential write" Transaction in C++

```
void arm::do_NonSeqW::body(tbvTaskArgBlock *args) {
// Body of do_NonSeqW
printf("\nEntering do_NonSeqW::body\n");
arg *myArgsP;
myArgsP = (arg *)args;
transStart(myArgsP);
printf("Initalize Cycle Type ...\n");
tbvWait(SIGNAL_D0);
*(this->parent_tvm_p->request_p) = 0ULL;
*(this->parent_tvm_p->seq_p) = 0ULL;
printf("Assert address and data size ...\n");
tbvWait(SIGNAL_D - SIGNAL_D0);
*(this->parent_tvm_p->address_p) = myArgsP->arg1;
*(this->parent_tvm_p->read_write_p) = 1ULL;
*(this->parent_tvm_p->mas_p) = myArgsP->arg2;
printf("Sends data and wait for slave nWAIT ...\n");
tbvWait(SIGNAL_D1 - SIGNAL_D);
*(this->parent_tvm_p->dreg_p) = myArgsP->arg3;
tbvWait(*(this->parent_tvm_p->clock_p),
(tbvThread::POSEDGE);
while (*(this->parent_tvm_p->wait_p) == 0ULL)
tbvWait(*(this->parent_tvm_p->clock_p),
tbvThread::POSEDGE);
printf("Terminate cycle sequentially ...\n");
tbvWait(SIGNAL_D0 - ASB_D);
*(this->parent_tvm_p->request_p) = 1ULL;
*(this->parent_tvm_p->seq_p) = 0ULL;
tbvWait(SIGNAL_D - SIGNAL_D0);
*(this->parent_tvm_p->address_p) = 0xffffffffULL;
*(this->parent_tvm_p->read_write_p) = 1ULL;
*(this->parent_tvm_p->mas_p) = 0x3ULL;
tbvWait(SIGNAL_D1 - SIGNAL_D);
*(this->parent_tvm_p->dreg_p) = "0xzzzzzzzz";
```

```
tbvWait(ASB_CYC - SIGNAL_D1);
*(this->parent_tvm_p->dreg_p) = "0xzzzzzzzz";
printf("Leaving do_NonSeqW::body\n");
transEnd(myArgsP);
}
```

Example 5-16 shows the test that incorporates the above transaction.

Example 5-16. Test Embedded with "non-sequential write" Transaction in C++

```
void arm_test() {
tbvTvm *arm_p;
myTaskArgBlockT myTaskArgs;
tbvTask *do_NonSeqW_p;
printf("\n Entering arm_test\n\n");
// Reference to TVM instance
arm_p = &tbvTvm::getTvmByInstanceName(
"EVOP_testbench.EVOP.FB.ARM.arm_inst");
do_NonSeqR_p = &arm_p
->getTaskByName("do_NonSeqR");
//Non Sequential write of data at 0x94000048ULL
myTaskArgs.arg1 = 0x94000048ULL;
myTaskArgs.arg2 = 0x2ULL;
myTaskArgs.arg3 = 0x12233445ULL;
do_NonSeqW_p->run(&myTaskArgs);
printf("time = %d ...\n", tf_gettime());
printf("\n Leaving arm_test\n\n");
printf("\n It works ...\n\n");
}
```

5.7.4 Transaction Analysis

The TBV environment has transaction recording and viewing capabilities. Both transaction-level and signal-level information can be recorded and analyzed. This enables easy debugging of the errors in the design. Each transaction is represented with a label and associated properties, such as transaction name, address, or data for read or write. TVMs can track, record, and display all transaction information, including errors.

Figure 5-14 shows an example display of the "non-sequential write" transaction `do_NotSeqw`. `mclk` is the bus clock signal, `nRW` is the read(0)/write(1) signal, `nwait` is the wait-state indication signal, `A` is the address bus, and `D` is the data bus. The transaction shows that the data "aabb5511" is being written in the memory address location "94000048."

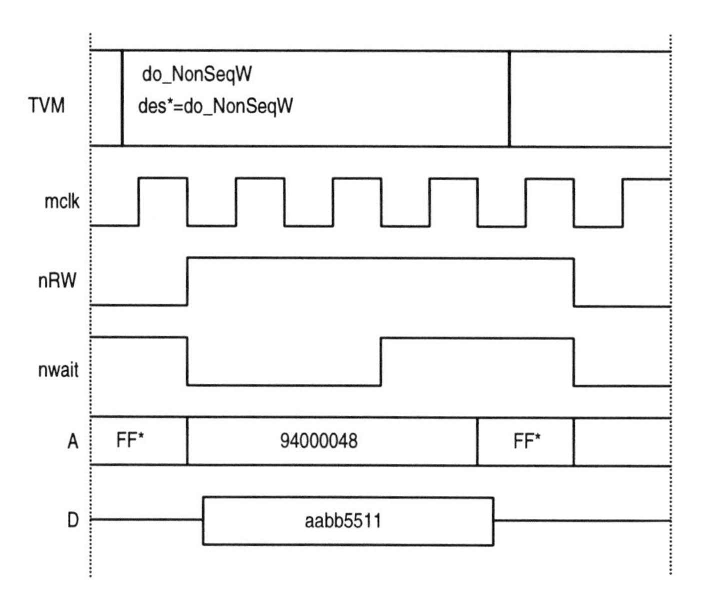

Figure 5-14. Non-sequential Write Transaction Waveforms

5.7.5 Function Coverage in TBV

Functional coverage assesses the degree to which functional features of a design have been exercised. Functional coverage data is generally a combination of temporal behavior (for example, a bus transaction) and the data associated with the behavior (for example, the transaction source or target). Functional coverage data can be obtained by cross-referencing functional coverage points (for example, the correlation of transactions on two different ports of a device or the correlation of processor instructions and processor interrupts). Unlike code coverage metrics, knowledge of the design is required to specify meaningful functional coverage metrics. The TBV tools are incorporated with features to report and analyze functional coverage on recorded transactions.

Figure 5-15 shows the functional coverage for an example design. The switched bar graph shows how many times the tx_rx transaction occurred, and the ports bar graph shows how many times the port was accessed.

Switched Bar Graph		Ports Bar Graph	
tx0_rx0	0	tx0	20
tx1_rx0	8	tx1	42
tx2_rx0	4	tx2	19
tx3_rx0	4	tx3	14
tx0_rx1	1	rx0	16
tx1_rx1	0	rx1	14
tx2_rx1	10	rx2	36

Figure 5-15. Functional Coverage Analysis at Transaction Level

5.7.6 TBV Methodology

TBV methodology assumes the availability of the gate-level netlist or RTL code for the DUT and the TVMs already created in other projects for reuse. As shown in Figure 5-16, the methodology steps are as follows:

5. **Check TVM library**: Check the design and search for suitable TVMs if available in the library. If the library does not contain the required TVMs, suitable TVMs should be created as per the design requirements.
6. **Create TVMs**: TVMs can be created in Verilog or C++, depending on the selection made by the team.
7. **Create test**: The test is created by embedding the tasks from the TVMs required to test the DUT.
8. **Compile and link**: Compile the files created and link them to run the simulation.
9. **Simulation**: Run the simulation on the RTL design using the tests created.

10. **Output analysis**: Analyze and debug the transaction response from the DUT for its intended functionality. Fix the errors found during the analysis.
11. **Functional coverage**: Check for the verification coverage.

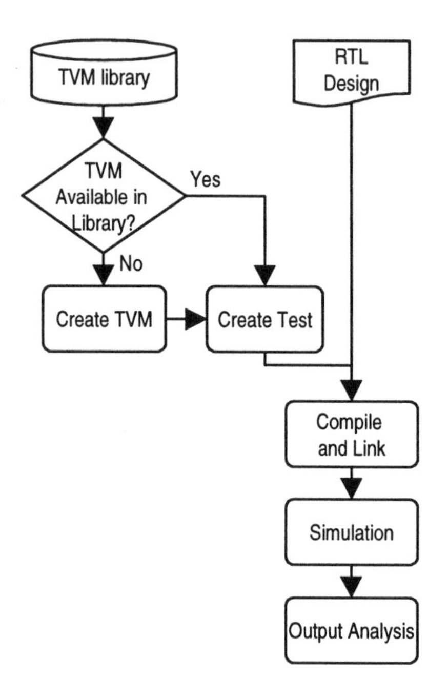

Figure 5-16. TBV Methodology Flow

5.7.7 Bluetooth SOC

Figure 5-17 shows a simple block diagram of the example Bluetooth SOC design. The ARM processor is replaced with a TVM to illustrate the TBV aspects. The remaining design is considered to be a DUT. The TVM is driven by a testbench. The testbench consists of a set of transactions created to test the functionality of the design.

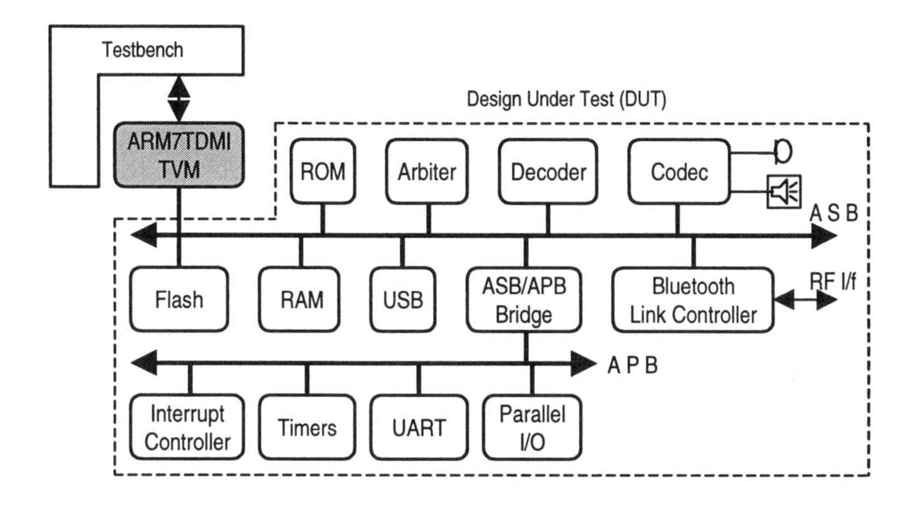

Figure 5-17. ARM Processor Replaced with TVM in Bluetooth SOC Design

5.7.7.1 Creating a TVM for ARM7TDMI

The first task in TBV is to identify the blocks that require the TVM. For example in the Bluetooth SOC, the ARM processor is replaced with a TVM. Example 5-17 shows the TVM for the ARM processor written in C++. It contains non-sequential read and write, and sequential read and write tasks for the ARM. The required header files are also given here.

Example 5-17. TVM for the ARM Processor

```
/*
* File: arm.cc
*/
#include "tb.h"
#include "arm.h"
#define ASB_CYC    20 // Cycle Time
#define ASB_D       5
#define HALF_ASB_CYC (ASB_CYC/2)
#define SIGNAL_D0  7
#define SIGNAL_D  13
#define SIGNAL_D1 18
//=======================================================
```

```
arm::arm() :
      clock(getFullInterfaceHdlNameP("clock")),
      reset(getFullInterfaceHdlNameP("reset")),
      request(getFullInterfaceHdlNameP("request")),
      wait(getFullInterfaceHdlNameP("wait")),
      seq(getFullInterfaceHdlNameP("seq")),
      address(getFullInterfaceHdlNameP("address")),
      dreg(getFullInterfaceHdlNameP("dreg")),
      mas(getFullInterfaceHdlNameP("mas")),
      read_write(getFullInterfaceHdlNameP("read_write"))
{
   printf("\n arm ctor called\n\n");
//-----------------------------------------------------
  // Tasks in this TVM
  do_NonSeqR_p = new do_NonSeqR(this);
  do_NonSeqW_p = new do_NonSeqW(this);
  do_SeqR_p =  new do_SeqR(this);
  do_SeqW_p =  new do_SeqW(this);
}
//=====================================================
arm::do_NonSeqR::do_NonSeqR(tbvTvmT *link) :
tbvTaskT(link,"do_NonSeqR") {
parent_tvm_p = (arm *)link;
}
//=====================================================

arm::do_NonSeqW::do_NonSeqW(tbvTvmT *link):
tbvTaskT(link,"do_NonSeqW") {
parent_tvm_p = (arm *)link;
}
//=====================================================
arm::do_SeqR::do_SeqR(tbvTvmT *link) : tbvTaskT(link,
"do_SeqR") {
parent_tvm_p = (arm *)link;
}
//=====================================================
arm::do_SeqW::do_SeqW(tbvTvmT *link) : tbvTaskT(link,
"do_SeqW") {
parent_tvm_p = (arm *)link;
}
//=====================================================
```

```
//This function is called from $tbv_connect to instanti-
ate this TVM.
void arm::arm_create() {
  new arm();
};
//===========================================================
void arm::do_NonSeqR::body(tbvTaskContextT *args) {
//  myArgBlockT  *myArgsP;
   printf("\nEntering do_NonSeqR::body\n");
  arg *myArgsP;
  myArgsP = (arg *)args;
  myArgsP->setCurrentFiber(parent_tvm_p->getFiberP());
  myArgsP->transBegin();
  printf("  arg1 = %x\n", myArgsP->arg1);fflush(stdout);
  printf("  arg2 = %x\n", myArgsP->arg2);fflush(stdout);
//currentTime = tf_gettime();
printf("Initializing cycle type at time  ...\n");
tbvWait(SIGNAL_D0);
parent_tvm_p->request = 0ULL;
parent_tvm_p->seq = 0ULL;
printf("Assert address and data size  ...\n");
tbvWait(SIGNAL_D - SIGNAL_D0);
parent_tvm_p->address = myArgsP->arg1; //addrs,
  //argument 1
parent_tvm_p->read_write = 0ULL;
parent_tvm_p->mas = myArgsP->arg2;    //whb, argument 2
printf("Release bus for slave to drive data  ...\n");
tbvWait(SIGNAL_D1 - SIGNAL_D);
parent_tvm_p->dreg = "0xzzzzzzzz"; //assign
   //32'hzzzzzzzz
printf("Read data when slave drives it  ...\n");
{tbvWaitCycle(parent_tvm_p->clock,
tbvThreadT::POSEDGE);}
while (parent_tvm_p->wait == 0ULL)
{tbvWaitCycle(parent_tvm_p->clock,
tbvThreadT::POSEDGE);}
printf("Terminate NonSeq Rd cycle ...\n");
tbvWait(SIGNAL_D0 - ASB_D);
parent_tvm_p->request = 1ULL;
parent_tvm_p->seq = 0ULL;
tbvWait(SIGNAL_D - SIGNAL_D0);
```

```
parent_tvm_p->address = 0xffffffffULL;  //addrs,
 //32'hffffffff;
parent_tvm_p->read_write = 0ULL;
parent_tvm_p->mas = 0x3ULL;       //whb, 2'b11
tbvWait(SIGNAL_D1 - SIGNAL_D);
parent_tvm_p->dreg = "0xzzzzzzzz";    //assign
 //32'hzzzzzzzz;
tbvWait(ASB_CYC - SIGNAL_D1);
parent_tvm_p->dreg = "0xzzzzzzzz";    //assign
32'hzzzzzzzz;
printf("Leaving do_NonSeqR::body\n");
myArgsP->transEnd();
}
//=======================================================
void arm::do_NonSeqW::body(tbvTaskContextT *args) {
  printf("\nEntering do_NonSeqW::body\n");
  arg *myArgsP;
  myArgsP = (arg *)args;
  myArgsP->setCurrentFiber(parent_tvm_p->getFiberP());
  myArgsP->transBegin();
  printf("  arg1 = %x\n", myArgsP->arg1);fflush(stdout);
  printf("  arg2 = %x\n", myArgsP->arg2);fflush(stdout);
  printf("  arg3 = %x\n", myArgsP->arg3);fflush(stdout);
//currentTime = tf_gettime();
printf("Initalize Cycle Type  ...\n");
tbvWait(SIGNAL_D0);
parent_tvm_p->request = 0ULL;
parent_tvm_p->seq = 0ULL;
printf("Assert address and data size  ...\n");
tbvWait(SIGNAL_D - SIGNAL_D0);
parent_tvm_p->address = myArgsP->arg1;//adrs,argument 1
parent_tvm_p->read_write = 1ULL;
parent_tvm_p->mas = myArgsP->arg2;//whb, argument 2
printf("Sends data and wait for slave nWAIT  ...\n");
tbvWait(SIGNAL_D1 - SIGNAL_D);
parent_tvm_p->dreg = myArgsP->arg3;//assign
                                   //32'hzzzzzzzz
{tbvWaitCycle(parent_tvm_p->clock,
tbvThreadT::POSEDGE);}
while (parent_tvm_p->wait == 0ULL)
{tbvWaitCycle(parent_tvm_p->clock,
```

```
tbvThreadT::POSEDGE);}
printf("Terminate cycle sequentially ...\n");
tbvWait(SIGNAL_D0 - ASB_D);
parent_tvm_p->request  = 1ULL;
parent_tvm_p->seq = 0ULL;
tbvWait(SIGNAL_D - SIGNAL_D0);
parent_tvm_p->address = 0xffffffffULL;//addrs,
                                      // 32'hffffffff;
parent_tvm_p->read_write = 1ULL;
parent_tvm_p->mas = 0x3ULL;  //whb, 2'b11
tbvWait(SIGNAL_D1 - SIGNAL_D);
parent_tvm_p->dreg = "0xzzzzzzzz";//assign
                                  // 32'hzzzzzzzz;
tbvWait(ASB_CYC - SIGNAL_D1);
parent_tvm_p->dreg = "0xzzzzzzzz";//assign
                                  // 32'hzzzzzzzz;
printf("Leaving do_NonSeqW::body\n");
myArgsP->transEnd();
}
//=======================================================
void arm::do_SeqR::body(tbvTaskContextT *args) {
printf("\nEntering do_SeqR::body\n");
arg *myArgsP;
myArgsP = (arg *)args;
  myArgsP->setCurrentFiber(parent_tvm_p->getFiberP());
  myArgsP->transBegin();
printf("  arg1 = %x\n", myArgsP->arg1);fflush(stdout);
printf("  arg2 = %x\n", myArgsP->arg2);fflush(stdout);
printf("Initialize Cycle Type  ...\n");
tbvWait(SIGNAL_D0);
parent_tvm_p->request = 0ULL;
parent_tvm_p->seq = 0ULL;
printf("Assert address for the words  ...\n");
tbvWait(SIGNAL_D - SIGNAL_D0);
parent_tvm_p->address = myArgsP->arg1;//addrs,
                                      // argument 1
parent_tvm_p->read_write = 0ULL;
parent_tvm_p->mas = 0x2ULL;
printf("Release bus for slave to drive data  ...\n");
tbvWait(SIGNAL_D1 - SIGNAL_D);
parent_tvm_p->dreg = "0xzzzzzzzz"; //assign
```

```
                                    // 32'hzzzzzzzz
printf("Sequentially increment address to read when
slave is ready ...\n");
{tbvWaitCycle(parent_tvm_p->clock,
tbvThreadT::POSEDGE);}
while (parent_tvm_p->wait == 0ULL)
{tbvWaitCycle(parent_tvm_p->clock,
tbvThreadT::POSEDGE);}
for (int i = 1; i <myArgsP->arg2; i = i +1)
{
        tbvWait(SIGNAL_D0 - ASB_D);
        parent_tvm_p->request  = 0ULL;
        parent_tvm_p->seq = 1ULL;
        tbvWait(SIGNAL_D - SIGNAL_D0);
        parent_tvm_p->address = myArgsP->arg1 + (i*4);
        {tbvWaitCycle(parent_tvm_p->clock,
          tbvThreadT::POSEDGE);}
        while (parent_tvm_p->wait == 0ULL)
        {tbvWaitCycle(parent_tvm_p->clock,
          tbvThreadT::POSEDGE);}
}
printf("Terminate Seq Rd cycle ...\n");
tbvWait(SIGNAL_D0 - ASB_D);
parent_tvm_p->request  = 1ULL;
parent_tvm_p->seq = 0ULL;
tbvWait(SIGNAL_D - SIGNAL_D0);
parent_tvm_p->address = 0xffffffffULL;//addrs,
                                       //32'hffffffff;
parent_tvm_p->read_write = 0ULL;
parent_tvm_p->mas = 0x3ULL;       //whb, 2'b11
tbvWait(SIGNAL_D1 - SIGNAL_D);
parent_tvm_p->dreg = "0xzzzzzzzz"; //assign
                                   // 32'hzzzzzzzz;
tbvWait(ASB_CYC - SIGNAL_D1);
parent_tvm_p->dreg = "0xzzzzzzzz"; //assign
                                   // 32'hzzzzzzzz;
printf("Leaving do_SeqR::body\n");
myArgsP->transEnd();
}
//====================================================
void arm::do_SeqW::body(tbvTaskContextT *args) {
```

```
  printf("\nEntering do_SeqW::body\n");
  arg *myArgsP;
  myArgsP = (arg *)args;
  myArgsP->setCurrentFiber(parent_tvm_p->getFiberP());
  myArgsP->transBegin();
  printf("  arg1 = %x\n", myArgsP->arg1);fflush(stdout);
  printf("  arg2 = %x\n", myArgsP->arg2);fflush(stdout);
  printf("  arg3 = %x\n", myArgsP->arg3);fflush(stdout);
if (myArgsP->arg3 == 0)
{
   printf("Begin of Sequential Write cycle  ...\n");
   tbvWait(SIGNAL_D0);
   parent_tvm_p->request = 0ULL;
   parent_tvm_p->seq = 1ULL;
   printf("Assert address and data size  ...\n");
   tbvWait(SIGNAL_D - SIGNAL_D0);
   parent_tvm_p->address = myArgsP->arg1;
   parent_tvm_p->read_write = 1ULL;
   parent_tvm_p->mas = 0x2ULL; //whb, argument 2
   printf("Sends data and wait for slave nWAIT  ...\n");
   tbvWait(SIGNAL_D1 - SIGNAL_D);
   parent_tvm_p->dreg = myArgsP->arg2;
   {tbvWaitCycle(parent_tvm_p->clock,
    tbvThreadT::POSEDGE);}
    while (parent_tvm_p->wait == 0ULL)
    {tbvWaitCycle(parent_tvm_p->clock,
      tbvThreadT::POSEDGE);}
}
else if (myArgsP->arg3 == 1)
{
  printf("Middle of Sequential Write cycle  ...\n");
  tbvWait(SIGNAL_D0 - ASB_D);
  parent_tvm_p->request = 0ULL;
   parent_tvm_p->seq = 1ULL;
   printf("Assert address and data size  ...\n");
   tbvWait(SIGNAL_D - SIGNAL_D0);
   parent_tvm_p->address = myArgsP->arg1;
   parent_tvm_p->read_write = 1ULL;
   parent_tvm_p->mas = 0x2ULL; //whb, argument 2
   printf("Sends data and wait for slave nWAIT  ...\n");
   tbvWait(SIGNAL_D1 - SIGNAL_D);
```

```
      parent_tvm_p->dreg = myArgsP->arg2;
      {tbvWaitCycle(parent_tvm_p->clock,
        tbvThreadT::POSEDGE);}
      while (parent_tvm_p->wait == 0ULL)
      {tbvWaitCycle(parent_tvm_p->clock,
        tbvThreadT::POSEDGE);}
   }
   else if (myArgsP->arg3 == 2)
   {
      printf("End of Sequential Write Cycle ...\n");
      printf("Initalize Cycle Type  ...\n");
      tbvWait(SIGNAL_D0 - ASB_D);
      parent_tvm_p->request = 0ULL;
      parent_tvm_p->seq = 1ULL;
      printf("Assert address and data size  ...\n");
      tbvWait(SIGNAL_D - SIGNAL_D0);
      parent_tvm_p->address = myArgsP->arg1;
      parent_tvm_p->read_write = 1ULL;
      parent_tvm_p->mas = 0x2ULL;
      printf("Sends data and wait for slave nWAIT  ...\n");
      tbvWait(SIGNAL_D1 - SIGNAL_D);
      parent_tvm_p->dreg = myArgsP->arg2;
      {tbvWaitCycle(parent_tvm_p->clock,
        tbvThreadT::POSEDGE);}
      while (parent_tvm_p->wait == 0ULL)
      {tbvWaitCycle(parent_tvm_p->clock,
        tbvThreadT::POSEDGE);}
      printf("Terminate cycle sequentially ...\n");
      tbvWait(SIGNAL_D0 - ASB_D);
      parent_tvm_p->request  = 1ULL;
      parent_tvm_p->seq = 0ULL;
      tbvWait(SIGNAL_D - SIGNAL_D0);
      parent_tvm_p->address = 0xffffffffULL;//addrs,
                                    // 32'hffffffff;
      parent_tvm_p->read_write = 1ULL;
      parent_tvm_p->mas = 0x3ULL;    //whb, 2'b11
      tbvWait(SIGNAL_D1 - SIGNAL_D);
      parent_tvm_p->dreg = "0xzzzzzzzz";//assign
                                 // 32'hzzzzzzzz;
      tbvWait(ASB_CYC - SIGNAL_D1);
      parent_tvm_p->dreg = "0xzzzzzzzz"; //assign
```

```
                                          // 32'hzzzzzzzz;
}

printf("Leaving do_SeqW::body\n");
myArgsP->transEnd();
}
```

The following header file defines all the signals used in the TVM.

```
arm.h

//Defines all the signals used.

#include "tb.h"
#include <stdio.h>
//====================================
class arm : public tbvTvmT { public:
// ctor:
arm();
// The interface signals:
tbvSignalHdlT clock;
tbvSignalHdlT reset;
tbvSignalHdlT request;
tbvSignalHdlT wait;
tbvSignalHdlT seq;
tbvSignalHdlT address;
tbvSignalHdlT dreg;
tbvSignalHdlT mas;

tbvSignalHdlT read_write;
//
tbvTaskT *do_NonSeqR_p;
tbvTaskT *do_NonSeqW_p;
tbvTaskT *do_SeqR_p;
tbvTaskT *do_SeqW_p;
//------------------------------------
// task Non Sequential Read:
```

```
class do_NonSeqR : public tbvTaskT {
 public:
do_NonSeqR(tbvTvmT *link);  // ctor
arm *parent_tvm_p;
virtual void body(tbvTaskContextT  *args);
};

//-----------------------------------------
// task Non Sequential Write:

class do_NonSeqW : public tbvTaskT {
public:
do_NonSeqW(tbvTvmT *link);  // ctor
arm *parent_tvm_p;
virtual void body(tbvTaskContextT  *args);

};

//-----------------------------------------
// task Sequential Read:

class do_SeqR : public tbvTaskT {
 public:
do_SeqR(tbvTvmT *link);  // ctor
arm *parent_tvm_p;
virtual void body(tbvTaskContextT  *args);

};

//-----------------------------------------
// task  Sequential Write:

class do_SeqW : public tbvTaskT {
 public:
do_SeqW(tbvTvmT *link);  // ctor
arm *parent_tvm_p;
virtual void body(tbvTaskContextT  *args);

};

//-----------------------------------------
```

```
// This function is called from $tbv_connect,to
// instantiate this TVM.
static void arm_create();

}; // class myTvm

//--------------------------------------
// The argument block type for this task:
struct myTaskArgBlockT {
  int arg1;
  int arg2;
};

class arg : public tbvTaskContextT
{
  public:

int arg1, arg2, arg3;

  protected:
  private:
};
```

All the TVMs are defined in a table in the arm_user.cc file.

```
/* All the TVMs are to be defined in a table in the
* 'arm_user.cc'  file.
* The first entry of an item in the table is the string
* name of the TVM that will be referenced from a
* $tbv_tvm_connect PLI call in the HDL.
* The second entry is the name of the user-written
* create function that will instantiate the TVM.
*/

#include "tb.h"
#include "arm.h"
tbvTvmTypeT tbvTvmTypes[] = {
{ "arm", arm::arm_create },
{ 0, 0}
```

```
};
extern void arm_test();
```

5.7.7.2 Creating a Test

A test is created by embedding the tasks from the TVM required to test the Bluetooth SOC design. Example 5-18 gives the test for the Bluetooth SOC design TBV.

Example 5-18. Test in C++ for Bluetooth SOC

```
arm_test.cc

#include <stdio.h>
#include "tb.h"
#include "arm.h"
extern "C" {int tf_gettime();}
void arm_test() {
  arm *arm_p;
//  myTaskArgBlockT myTaskArgs;
  arg myTaskArgs;
  arm::do_NonSeqR *do_NonSeqR_p;
  arm::do_NonSeqW *do_NonSeqW_p;
  arm::do_SeqR *do_SeqR_p;
  arm::do_SeqW *do_SeqW_p;
  printf("\n  Entering arm_test\n\n");
  arm_p = (arm *)tbvTvmT::getTvmBy-
FullNameP("EVOP_testbench.EVOP.FB.ARM.arm_inst");
  if (arm_p == 0) {
    printf("*** TVM not found\n");
  } else {    printf("    Found the TVM\n");  }
  do_NonSeqR_p = (arm::do_NonSeqR *)arm_p->getTask-
ByNameP("do_NonSeqR");
  do_NonSeqW_p = (arm::do_NonSeqW *)arm_p->getTask-
ByNameP("do_NonSeqW");
  do_SeqR_p = (arm::do_SeqR *)arm_p->getTask-
ByNameP("do_SeqR");
  do_SeqW_p = (arm::do_SeqW *)arm_p->getTask-
ByNameP("do_SeqW");
  if (do_NonSeqR_p == 0) { printf("*** Task do_NonSeqR
not found\n"); }
```

```
else {    printf("Task do_NonSeqR found\n");  }
  if (do_NonSeqW_p == 0) {
    printf("*** Task do_NonSeqW_p not found\n");  }
else {    printf("Task do_NonSeqW_p found\n");  }
  if (do_SeqR_p == 0) {    printf("*** Task do_SeqR_p
not found\n");  }
else {    printf("Task do_SeqR_p found\n");  }
  if (do_SeqW_p == 0) { printf("*** Task do_SeqW_p not
found\n");}
else { printf("Task do_SeqW_p found\n");}
//If you want to be able to change the runs to spawns,
create a duplicate
//function for the arg class, and in the constructor,
call the method
//setAutoCopy(true).
//Sequnetially read the data starting at 94000040
  // Call a task in the TVM
  myTaskArgs.arg1 = 0x94000040ULL;
  myTaskArgs.arg2 = 0x4ULL;
  do_SeqR_p->run(&myTaskArgs);
printf("time = %d ...\n", tf_gettime());
//Non Sequential Read of data
  myTaskArgs.arg2 = 0x2ULL;
  printf("Calling the task\n");
  do_NonSeqR_p->run(&myTaskArgs);
  myTaskArgs.arg1 = 0x94000044ULL;
  myTaskArgs.arg2 = 0x2ULL;
  do_NonSeqR_p->run(&myTaskArgs);
//Non Sequential write of data at 0x94000048ULL
  myTaskArgs.arg1 = 0x94000048ULL;
  myTaskArgs.arg2 = 0x2ULL;
  myTaskArgs.arg3 = 0x12233445ULL;
  do_NonSeqW_p->run(&myTaskArgs);
 printf("time = %d ...\n", tf_gettime());
//Non Sequential Read of data from 0x94000048ULL
  myTaskArgs.arg1 = 0x94000048ULL;
  myTaskArgs.arg2 = 0x2ULL;
  do_NonSeqR_p->run(&myTaskArgs);
//Sequentail write of data starting at 0x94000020ULL
  myTaskArgs.arg1 = 0x94000020ULL;
  myTaskArgs.arg2 = 0xcccc1111ULL;
```

```
  myTaskArgs.arg3 = 0x0ULL;
  do_SeqW_p->run(&myTaskArgs);
  myTaskArgs.arg1 = 0x94000024ULL;
  myTaskArgs.arg2 = 0x99991111ULL;
  myTaskArgs.arg3 = 0x1ULL;
  do_SeqW_p->run(&myTaskArgs);
  myTaskArgs.arg1 = 0x94000028ULL;
  myTaskArgs.arg2 = 0x88881111ULL;
  myTaskArgs.arg3 = 0x1ULL;
  do_SeqW_p->run(&myTaskArgs);
  myTaskArgs.arg1 = 0x9400002cULL;
  myTaskArgs.arg2 = 0x55554444ULL;
  myTaskArgs.arg3 = 0x2ULL;
  do_SeqW_p->run(&myTaskArgs);
//Sequential read of data stored above
  myTaskArgs.arg1 = 0x94000020ULL;
  myTaskArgs.arg2 = 0x4ULL;
  do_SeqR_p->run(&myTaskArgs);
  printf("\n  Leaving arm_test\n\n");
  printf("\n  It works ...\n\n");
}
void tbvMain() {
  arm_test();
}
```

5.7.7.3 Compilation and Simulation

All the files created for TVM and test are compiled along with the design, and the simulation is run. After simulation, the TBV tool generates a log file that can be examined. The transactions can also be checked for correct functionality. Example 5-19 shows a log file that was obtained after running a simulation on the Bluetooth SOC design.

Example 5-19. Log File for the Bluetooth SOC Design

```
 Entering arm_test
    Found the TVM
Task do_NonSeqR found
Task do_NonSeqW_p found
Task do_SeqR_p found
Task do_SeqW_p found
```

```
Entering do_SeqR::body
  arg1 = 94000040
  arg2 = 4
Initialize Cycle Type  ...
Assert address for the words  ...
Release bus for slave to drive data  ...
Sequentially increment address to read when slave is
ready ...
Terminate Seq Rd cycle ...
Leaving do_SeqR::body
time = 0 ...
Calling the task
:
:
:
...............
  Leaving arm_test
  It works ...
Memory Usage - 7.9M program + 29.0M data = 36.9M total
CPU Usage - 0.4s system + 1.2s user = 1.6s total (79.5%
cpu)
Simulation complete via $finish(2) at time 12200 NS + 0
$
finish(2);
ncsim> exit
```

5.8 Simulation Acceleration

Simulation accelerators address the speed problem of software simulators by providing a dedicated execution platform for the simulator application and the HDL design. The dedicated execution platform can be an emulation box, a hardware accelerator, or a rapid prototyping system. The simulation speed obtainable can be over 1 MHz, depending on the method of acceleration chosen.

5.8.1 Emulation

Emulation is a technology in the world of design verification that has become increasingly popular in the past few years. This technology has been used successfully on a number of complex design projects. It involves mapping the DUT into a reconfigurable hardware platform built from array processors or field programmable gate array (FPGA) devices. Emulation systems can deliver very high simulation performance.

Some of the features of emulation are:

- Enables the early creation of a hardware model of the chip
- Enables the user to detect and correct the bugs early in the design
- Ability to develop and debug the application software
- Handles design complexity of 50,000 to 20 million gates
- Runs at near real-time speeds in some cases
- Reconfigurable for a variety of applications
- Scalable as per the design complexity and cost
- Ability to connect in-circuit emulators (ICE)
- In-built logic analyzer facility for debugging
- Probes and monitors the pin/signal in the system
- Bus models for standard buses, such as peripheral component interconnect (PCI), Ethernet, and others

The emulation solutions available in the industry are array processor-based and FPGA-based.

5.8.1.1 Array Processor-based Emulation

Array processor-based emulation consists of an array of high-speed processors and large high-speed, multiported memory fitted into a system or box. The system is connected with the standard workstation. Figure 5-18 shows a block diagram of an array processor-based emulation environment.

In this method, the RTL design is compiled and downloaded to the emulation system for execution. The simulation speeds obtained from the system depend on the number of processors running concurrently. The system provides ICE capability, enabling the user to debug the design in the target system environment. A facility to monitor the signals through the logic analyzer is also incorporated in the system. The system provides an interface to CBS tools for easy migration between the environments.

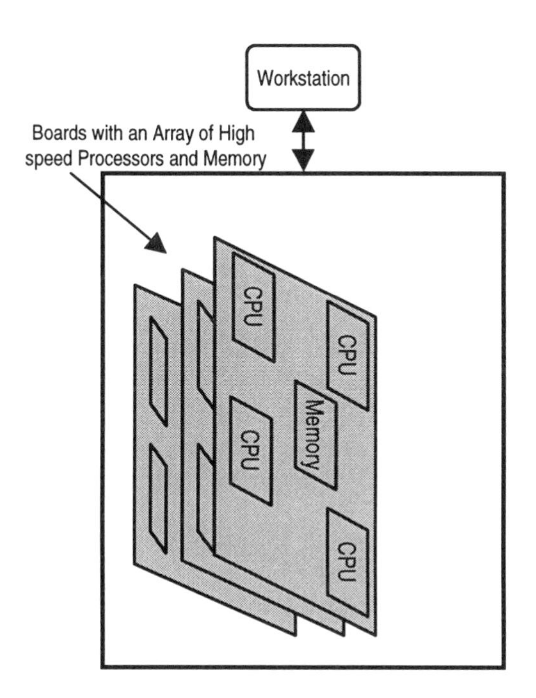

Figure 5-18. Block Diagram of an Array Processor-based Emulation System

Some of the emulation systems have multiuser capability. In large designs, the design can be partitioned and run on the emulation box to complete the verification faster.

5.8.1.2 FPGA-based Emulation

FPGA-based emulation consists of interconnected, high-speed, and high-density FPGAs. The system can incorporate custom integrated circuits (IC), memory ICs, bonded-out cores, and discrete components. Figure 5-19 shows a simple block diagram of an FPGA- based emulation system.

The system can be configured with the number of FPGAs required. Most systems provide programmable interconnection among the FPGAs through programmable electronic cross-bar switches. Also, some systems provide a facility to connect external modules so that it can interface circuit parts that cannot be incorporated within the system. The external modules include data conversion interfaces, such as analog-to-digital converters (ADC), digital-to-analog converters (DAC), analog filters, and radio-frequency (RF) modules.

FPGA-based emulation systems accept the RTL code of the design and testbench as input. It is preferable to create a synthesizable testbench to obtain high performance. In the case where the testbench is behavioral and not synthesizable, standard software simulators are used to run the testbench. Debugging uses ICE, and analysis uses logic analysis features incorporated in the system.

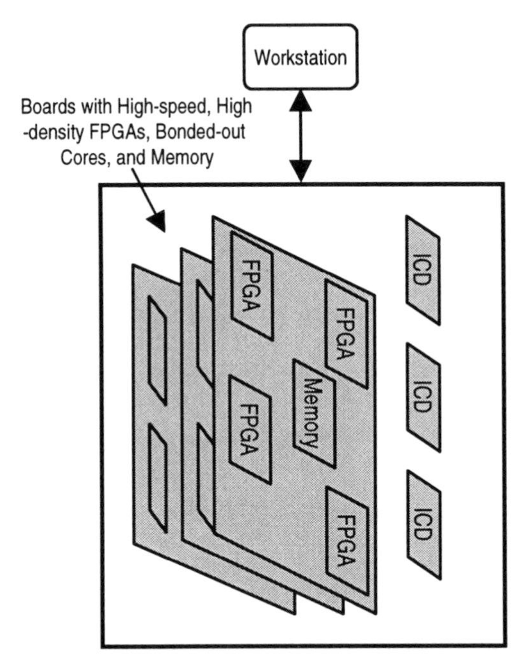

Figure 5-19. Block Diagram of an FPGA-based Emulation System

5.8.2 When to Use Emulation

Emulation can be used in the following situations prior to chip design tape-out.

- **Speed up simulation**: The emulation solutions available in the industry provide a very high simulation performance. Emulation requires a large number of functional test vectors (in the order of trillions). The typical number of functional test vectors (in the order of millions) used for normal simulation using software simulators is not enough to call for use of emulation. The vectors required for the emulation can be best provided by executing application programs and by executing random instruction sequences. This will provide overall improved coverage.

- **Increase verification cycles**: Emulation allows increased verification cycles, thereby building a high level of confidence in the design functionality.
- **Test real applications**: With some emulation systems, it is possible to develop, debug, and benchmark the real application software prior to tape-out.

To use an emulation in a project, the planning should begin right at the start, and it should be included in the verification plan. This helps in the following areas:

- Identifying the right emulation choice and avoiding unknown problems that could arise later
- Determining design modeling issues
- Identifying the emulation team, solution provider, and consulting group
- Leveraging the emulation work done earlier for other projects within the group or company

5.8.3 Emulation Environment

The emulation environment consists of the following elements, as shown in Figure 5-20.

- **Design input**: Accepts RTL code or gate-level netlist as the input. RTL code can be in Verilog, VHDL, or mixed language (mixed Verilog and VHDL code).
- **Testbench**: Most emulation systems accept synthesizable testbenches. The testbench generated for CBS works with few or no modifications. Some emulation systems accept behavioral testbenches. When a behavioral testbench is used, it is run on normal software simulators, and the synthesizable portion is run in the emulation system after synthesis.
- **Workstation**: A workstation is used as a host machine to interface with the emulation system. It is used to download, compile, and transfer the design data to the emulation system as well as control and debug the design. The compiler maps the DUT to a suitable form as required by the emulation system. The software simulator runs on the workstation with the design's behavioral testbench.

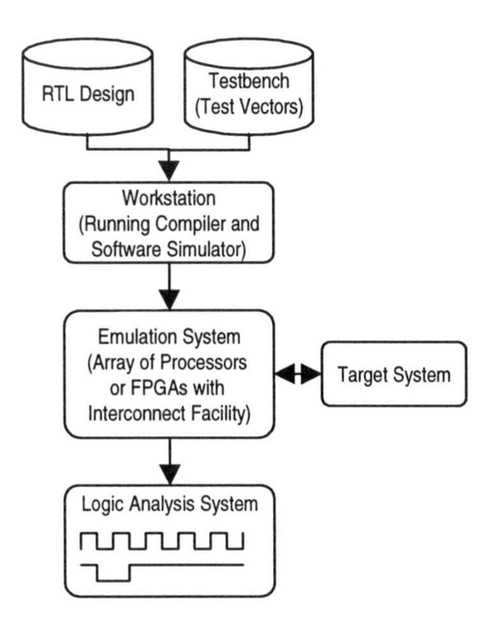

Figure 5-20. Block Diagram of an Emulation Environment

- **Emulation system (hardware platform)**: Consists of an array of programmable devices that are interconnected together. The programmable devices can be high-speed processors, or high-speed and high-density FPGAs. The interconnect can be fixed or programmable, depending on the design of the emulator.
- **Analysis**: Design analysis uses the facility provided in the emulation solution. In some emulation systems, an external logic analyzer is used; some provide in-built logic analysis capability.
- **Debugging**: Design debugging uses the facility provided in the emulation solution. In some emulation systems, an external ICE hardware module is used; some provide a built-in ICE.

5.8.4 Selecting an Emulation Solution

Some of the aspects to consider when selecting an emulation solution are:

- **Design capacity**: Check the design handling capacity of the environment. The vendors claim capacities of over 20 million gates.

- **Emulation speed**: The emulation speed available vary. Check for the benchmarking obtainable by having your own design models run on the emulation system. Some vendors claim emulation speeds of over 1 MHz.
- **Design input**: The design input varies from vendor to vendor. Some emulators accept RTL code described in HDL, such as Verilog or VHDL. Some require gate-level netlist, and some require the user to partition the design as per the requirement of the emulation solution.
- **Analysis and debug features**: Check the data capture and logic analysis capabilities, which should provide an easy way of locating errors and debugging.
- **Engineering change order** (ECO): An easy way of modifying the design when required should be available.
- **Schedule**: Check the time required to bring up the emulation environment. This may seriously affect the overall project schedule.
- **Cost**: Depends on design handling capacity, speed, and environment requirements. The current rate for an emulation solution is about $1 per gate. It should be noted that the interpretation of gate varies from vendor to vendor.
- **Support**: Check for complete and usable documentation, technical support, the ability to download upgrades, and training.
- **Usage for future projects**: Since the investment required in establishing an emulation environment is considerably high, find out how the solution can be used for future projects in order to leverage the investment.
- **Vendor reputation**: The reputation, long-term stability, and reliability of the product and vendor is an important issue. A reference check with the present users of the solution under consideration will help understand and give added confidence in the decision-making process.

5.8.5 Limitations of Emulation

Some of the limitations of emulation technology are:

- **Technology issues**: Only takes care of the design's functionality at clock-cycle boundaries. It does not address any timing and chip manufacturing technology-related problems. Emulation does not eliminate the need for timing analysis.
- **X and Z States**: The logic circuits can have 0,1, X, and Z states. The emulation takes care of 0, 1 states only, and X, Z states are ignored. This affects the initialization of the design.

- **System interaction**: Unlike software simulators, some hardware emulators do not support the interactive means of stopping the simulation whenever required to check the nets, modify, and restart the simulation.
- **Skill set**: Usually requires a different skill set and a rethinking of normal design and verification methodology, in addition to a large capital investment in hardware.

5.8.6 Emulation Methodology

The methodology steps are as follows, as shown in Figure 5-21.

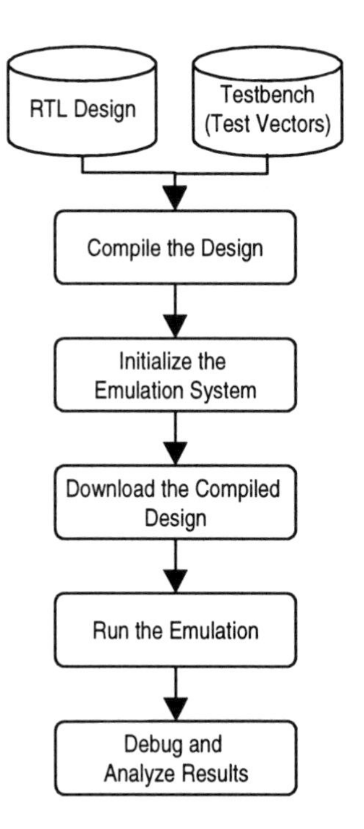

Figure 5-21. Emulation Methodology Flow

1. **Design input**: Accepts as input gate-level netlist or RTL code in Verilog, VHDL, or mixed-level.
2. **Testbench creation**: Can be synthesizable or behavioral.

3. **Compile**: Uses the compiler available on the workstation. The compiler is a very critical component of the emulation system, since a considerable time is spent in compiling the design.
4. **Initialize and download**: Initialize the emulation system and download the compiled design onto the system.
5. **Run emulation**: If the testbench is available in behavioral level, it is run on the standard software simulator, and the design with synthesizable testbench is run on the emulation system.
6. **Debug and analysis**: Design debugging uses the ICE facility, and analysis uses the logic analysis feature available in the emulation system. In the caes where these facilities are not available, an external ICE and logic analyzer are connected to the ports available on the emulation system.

5.8.7 Rapid Prototyping Systems

Rapid prototype systems are hardware design representations of the design being verified. The key to successful rapid prototyping is to quickly realize the prototype. Some approaches include emulation, as discussed above, reconfigurable prototyping systems, in which the target design is mapped to off-the-shelf devices, such as control processors, DSPs, bonded-out cores, and FPGAs. These components are mounted on daughter boards, which plug into a system interconnect motherboard containing custom programmable interconnect devices that model the target system interconnect. Application-specific prototypes map the target design to commercially available components and have limited expansion and reuse capability. Typically, these prototypes are built around board support packages (BSPs) for the embedded processors, with additional components (memories, FPGAs, and cores) added as needed.

5.8.8 Hardware Accelerators

A hardware accelerator is a platform that can be used to execute a complete design or a portion of the design. In most of the solutions available in the industry, the testbench remains running in software, while the actual design being verified is run in the hardware accelerator. Some of the solutions also provide acceleration capabilities for the testbenches.

5.8.9 Design Partitioning

Design partitioning can accelerate the design in one of two ways:

- Partitioning the design into several functional blocks. The system-level block test is extracted from an abstract model of the design running the full system testbench. The individual blocks can then be verified in isolation, with their associated system testbench.
- Running system simulations in a mixed-level mode where most blocks are run with abstract models of the design, and the detailed design is substituted for each block in turn.

Summary

A variety of simulation techniques and tools are available for speeding up the functional simulation process. As the time and cost to verify SOC designs increase, more efficient simulation methodologies need to be adopted.

References

1. Riches Stuart, Abrahams Martin. Practical approach to improving ASIC verification efficiency, Integrated System Design, July 1998.

2. Swami Ravi, Mandava Babu. Interconnect verification forms the linchpin of a DSL VDSM design, Integrated System Design, March 1999.

3. Averill R M, Barkley K G, ... Chip integration methodology for the IBM S/390 G5 and G6 custom microprocessors, Journal of research and development, Vol. 43, No. 5/6 - IBM S/390 Server G5/G6, 1999.

4. Cycle-based verification, a technical paper, www.quickturn.com.

5. SpeedSim vs. special accelerators, a technical paper, www.quickturn.com.

6. Bassak Gil. Focus report: HDL simulators, Integrated System Design, June 1998.

7. Cox Steve. Expose design bugs with transaction-based verification, Electronics Engineer, July 1998.

8. User manuals for Cadence Transaction-based verification tools, SpeedSim, and NC-Verilog.

9. EDA Watch, System chip verification: Moving from ASIC-out to System-In methodologies, Electronic Design, November 3, 1997.

10. ASIC Design methodology primer, ASIC Products application note, www.chips.ibm.com.

11. Doerre George, Colbourne Richard, ... Using advanced design methodology with first-time-success to speed time-to-market, IBM-MicroNews, First quarter 2000, Vol. 6, No.1.

12. Sullivan Janine. Design verification tools: A buyer's guide, Communication System Design, June 2000, www.csdmag.com.

13. Singletary Alan. Run it first, then build it - Core emulation in IBM microelectronics, IBM-MicroNews, Vol. 4, No. 1.

14. Balph Tom, Li Wilson. Hardware emulation accelerates HDL functional verification, Integrated System Design, April 1998.

15. Emulation, a technical paper, www.quickturn.com.

16. CoBALTplus, Technical data sheet, www.quickturn.com.

17. Mercury, Technical data sheet, www.quickturn.com.

18. Xcite-1000, Data sheet, www.axiscorp.com.

19. Raam Michael, Accelerator speeds HW/SW co-verification, EE Times, 6/5/00, www.eetimes.com.

20. Shieh Eric. Reconfigurable computing acclerates verification, Integrated System Design, January 2000.

21. SimExpress Hardware emulator data sheet, www.mentor.com.

22. Celaro: State-of-the-art Hardware emulator, www.mentor.com.

23. Tuck Barbara. Emulation steps up to verify complex designs, Computer Design, February 1997.

24. NSIM data sheets, www.ikos.com.

CHAPTER 6

Hardware/ Software Co-verification

To verify hardware (HW) and debug software (SW) running in a highly integrated system-on-a-chip (SOC) poses engineering challenges. The processor cores embedded in the SOC are no longer visible, since there are no pins available to connect an in-circuit emulator (ICE) and logic analyzer (LA) for debugging and analysis. An ICE and LA require address, data, and control buses for debugging, but these signals are hidden in the SOC. In addition to the functional verification of the hardware, the methodology must take into account the increasing amount of software being used in consumer electronic products.This chapter addresses the following topics:

- HW/SW co-verification environment and methods
- Soft prototypes
- Co-verification
- Rapid prototype systems
- FPGA-based design
- Developing printed circuit boards
- Software testing

The soft prototype and HW/SW co-verification methodologies are illustrated with an example of debugging the universal asynchronous receiver and transmitter (UART) device driver used in the Bluetooth SOC design.

6.1 HW/SW Co-verification Environment

In the SOC system design cycle, an abstract model of the design is created and simulated. This abstract functionality is then mapped to a detailed architecture of the system, and architectural performance modeling is performed. Architectural mapping partitions the design into hardware and software components, and the specifications are handed off to the hardware and software teams for implementation. The hardware team implements the hardware portion of the design in Verilog or VHDL, using hardware simulators for verification. The software team codes the software modules in assembly, C, or C++ languages and uses processor models or ICEs to test the software. Traditionally, the software team then waits for a hardware prototype for the final system integration.

Many problems can arise during the system integration process. The problems are due to such things as misunderstanding specifications, improper interface definitions, and late design changes. Errors can be eliminated with work-arounds in software, which may affect system performance, or with hardware modifications, which can be very costly and time-consuming, especially if it involves recycling an integrated circuit (IC). Moving the system integration phase forward in the design cycle would help in detecting these integration problems earlier. This can be achieved by creating a HW/SW co-verification environment early in the design cycle.

Some of the areas that are important for a HW/SW co-verification environment are:

- **Accuracy**: Models used in the environment should be cycle- or pin-accurate and mapped to the SOC functionality.
- **Performance**: The environment should be fast enough to run the software containing the real-time operating system (RTOS) and application.
- **Usage**: Both the hardware and software teams should be able to use the environment for functional and performance verification.
- **Availability**: To meet time-to-market goals and to enable HW/SW co-design and co-verification, the environment should be available early in the design cycle.
- **Cost**: Depends on the environment method that is being considered as well as the accuracy, performance, number of users, and the configuration required.

Figure 6-1 shows a simple block diagram of the HW/SW co-verification environment. The steps vary according to the method used.

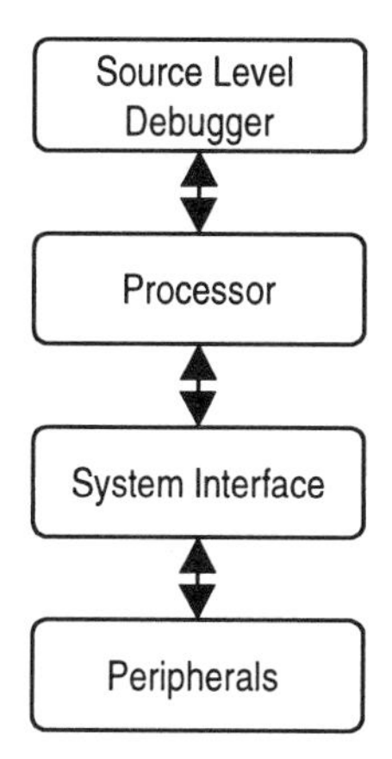

Figure 6-1. HW/SW Co-verification Environment

The environment consists of a source-level debugger that allows the user to download the firmware/software and interface with the system. The debugger helps to read, set breakpoints, reset, halt, and control the execution of the processor. Depending on the environment method, the processor block can be replaced with a simulation model, such as an instruction set simulator (ISS), a bus function model (BFM), register-transfer level (RTL) code, or a processor bonded-out core. The remainder of the system (the peripherals) are represented by C models, RTL code, real chips, or implemented in field programmable gate arrays (FPGA). The system interface provides the interface between the processor and the peripherals.

The methods described in this chapter use this basic environment. They support early integration of the hardware and software and overcomes the problems associated with traditional integration and verification methods.

The software porting and testing done in the HW/SW co-verification environment can be directly used for both silicon bring-up, assisting with quick validation of the design and product.

6.2 Emulation

Emulation has been used successfully on a number of complex design projects. It involves mapping the design under test (DUT) into a reconfigurable hardware platform built with an array of processors or FPGA devices. Emulation systems can deliver very high simulation performance.

Figure 6-2 shows a simple block diagram of an emulation environment. The processor, system interface, and peripherals are incorporated in the emulation system. The emulator interfaces with the source-level debugger through an ICE hardware module.

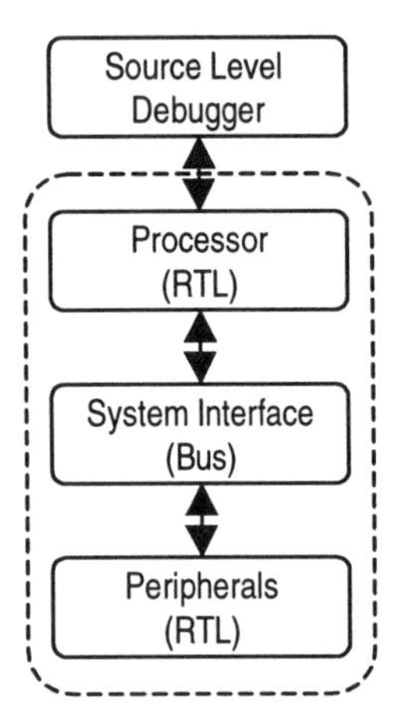

Figure 6-2. Emulation Environment

Emulation is covered in more detail in Chapter 5, "Simulation."

6.3 Soft or Virtual Prototypes

A soft or virtual prototype is a software design representation of the design being verified. It allows the activity of the processor registers, memory accesses, and peripherals to be checked. It is possible to run the actual application software and firmware if the host machine used to run the soft prototype is fast enough. This allows designers to make trade-offs by modifying system parameters and checking the results prior to the availability of actual silicon.

Figure 6-3 shows a simple block diagram of the soft prototype environment. The processor is modelled by an ISS that interfaces with the source-level debugger. The peripherals are represented with C-models.

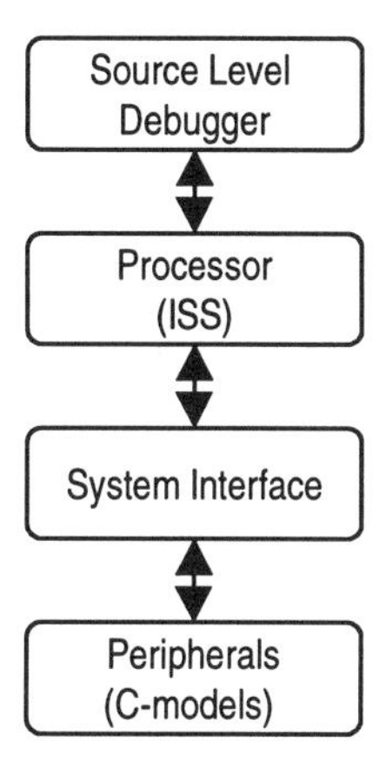

Figure 6-3. Soft Prototype Environment

The soft prototype allows designers to do the following:

- Make trade-offs by modifying system parameters and checking the results
- Test interrupt handlers
- Develop and test device drivers
- Test the correctness of compiler-generated code
- Visualize the behavior of the system and peripherals
- Test the correctness of the application algorithms

Figure 6-4 shows a simple block diagram of a typical soft prototype for the Bluetooth SOC design. It consists of a CPU debugger (for example, the ARM Debugger), a CPU ISS (for example, the ARMulator), and C models of all the peripherals. The software/firmware developed using the soft prototype can be reconfigured for emulation and downloaded through an ICE to the rapid prototype or target hardware system for testing.

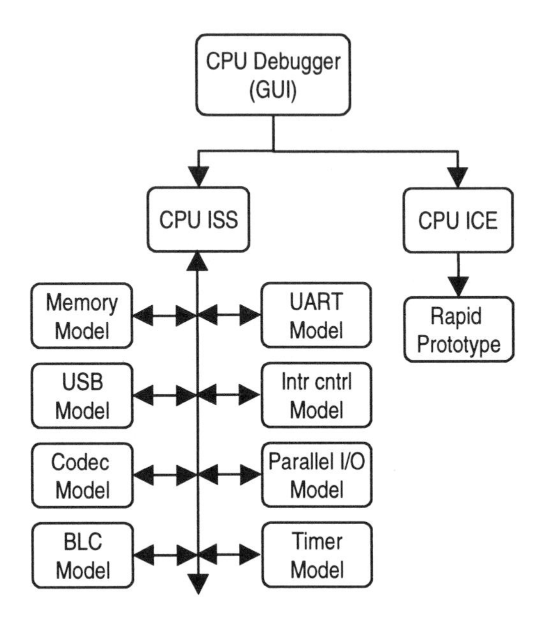

Figure 6-4. Soft or Virtual Prototype

To build a soft prototype of a processor-based system, the following software components are required.

- **Processor debugger**: It is a software module that enables the user to control the state of the debugging session through a target system. It is an integrated part of a debug kernel and user interface software. The debug kernel is integrated with the host machine that is used for debugging, the compiler, and the linker toolset. The debugger can be connected to a processor in the target hardware system through an ICE. The hardware can also be simulated through a software simulator, such as an ISS. The debugger can be used to read/write the registers or memory, halt, control execution, and restart the processor in a system or simulator.
- **Instruction set simulator**: It simulates the processor instructions without timing considerations. Some of the processor vendors provide an ISS. For example, the ARMulator offered by ARM emulates the ARM family of processors, provides the capability to interface the peripheral C models, and helps in creating a soft prototype.

- **C models of peripherals**: All the details of the registers, bit definitions, and interrupt behavior of the peripherals should be included in the C model. This helps for using the firmware/software in real silicon with no or few modifications.

6.3.1 Limitations

The soft prototype has the following limitations.

- **Limited capacity**: Because of limited capacity, it is restricted to early testing, interface debugging, and code segment debugging.
- **Limited speed**: In most cases, the speed with which the simulation can operate will become a critical factor, since simulators can never run as fast as the real processor.
- **Accuracy of models**: The peripheral models are functionally correct but not cycle-accurate and pin-accurate.
- **Synchronization**: It is often difficult to resolve the synchronization requirements of the peripheral data dependencies.

6.3.2 Methodology for Creating a Soft Prototype

The methodology steps to create a soft prototype are as follows, as shown in Figure 6-5.

1. **Study ISS features**: Study the ISS features of the processor that is selected for the design.
2. **Check interface support**: Check whether the ISS for the selected processor has the capability to interface the external C models to be used. If the interface support is not available in the ISS, a soft prototype cannot be created.
3. **Create C models**: Study the interface, memory address, registers, bit definitions, and interrupt behavior details of the peripherals and model them in C language.
4. **Write application code**: Write the application code that is to be verified using the soft prototype.
5. **Compile**: Compile the ISS, the C models of the peripherals, and the application software and generate an executable to be run on the host machine. The host machine can be a personal computer or a workstation.

6. **Run the executable**: Run the compiled executable program with the debugger. Start debugging. If there are any errors, fix them accordingly. The errors may be due to incorrect modeling of the peripherals or errors in application software.
7. **Test software**: Run in co-verification, emulation, or rapid prototyping environments for performance testing. The software can also be used for the final hardware and software integration when the silicon is available.

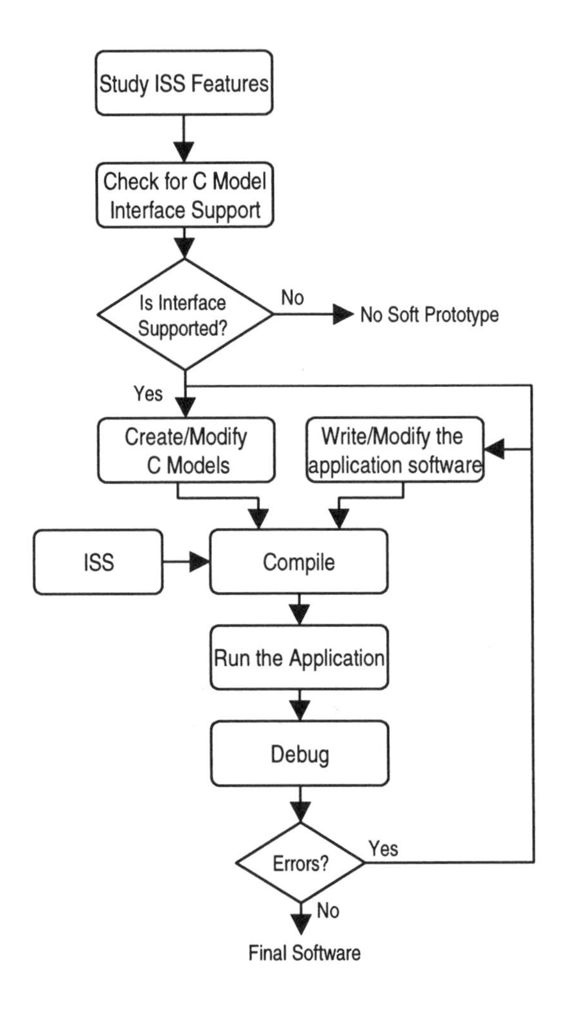

Figure 6-5. Methodology for Creating a Soft Prototype

6.3.3 Soft Prototype for the Bluetooth SOC

The soft prototype for the Bluetooth SOC design is created using the guidelines in the "Application note 32. The ARMulator" (refer to www.arm.com). Figure 6-6 shows the block diagram of the Bluetooth SOC design. The blocks used for the soft prototype are highlighted.

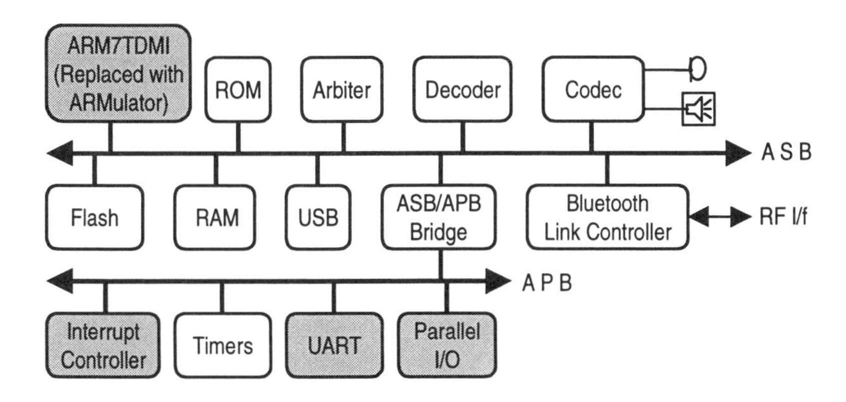

Figure 6-6. Soft Prototype of Bluetooth SOC

The C models are created for the UART, parallel port interface, and interrupt controller. The ARM7TDMI processor is replaced with the ARMulator, which consists of an ARM processor core model, memory interface, operating system interface for the execution environment, and a coprocessor interface. The ARMulator also interfaces with the ARM debugger.

6.3.3.1 Adding Peripheral Models

This section discusses adding the UART, parallel port interface, and interrupt controller models to the existing ARMulator memory model.

The `MemAccess` function in the copy of the ARMulator source code is modified to emulate a simulation environment in which devices connected to the advanced peripheral bus (APB) interrupt the processor through the interrupt controller.

The data is read into the UART receive buffer from a file. The UART interrupts the processor after reading a byte from the file. When the UART is ready to transmit the data, it interrupts the processor, reads the data, and shifts it into the transmit buffer.

The following steps are required to add peripheral models to the ARMulator. The steps are indicated in Example 6-1.

Step 1: Copy `armflat.c` to `Uart_pi_intc.c`. Include `stdio.h` in the file.

Step 2: Find the structure definition for `toplevel` in `Uart_pi_intc.c` and add the pointers `int UART_IN_IRQ` and `int UART_OUT_IRQ`, and file handles `uart_infile` and `uart_outfile`. `UART_IN_IRQ` and `UART_OUT_IRQ` are used to indicate that the interrupt from the UART is generated. `uart_infile` and `uart_outfile` are file handles for the UART input and output files.

Step 3: Modify `#define ModelName (tag_t)` to include the model name:

```
#define ModelName (tag_t) "UART_PI_INTC"
```

Step 4: Modify `armul.cnf` so that the default memory model reflects name of the memory model:

```
;Default=Flat becomes
 Default=UART_PI_INTC
```

Step 5: Create new functions (`Common_Irq` for the interrupt handler, `pi_set_irq` for the parallel port, `uart_set_in_irq` for the UART input interrupt, `uart_set_out_irq` for the UART output interrupt), which will be called by the `ARMulSchedule_Event` function.These functions trigger the respective IRQ.

Step 6: Modify the definition of `ARMul_MemStub` in `Uart_pi_intc.c` to reflect the name of the memory model:

```
ARMul_MemStub UART_PI_INTC = {
                              MemInit,
                                ModelName

                               }
```

Step 7: At the end of the MemInit function, add instructions to copy the ARMul_State pointer to the toplevel->state structure member. Also, initialize the other new structure members.

Step 8: In the models.h file, insert the new memory model stub entry after MEMORY{ARMul_MapFile}:

MEMORY{UART_INTC}

Step 9: Write a code for the interrupt controller registers, the UART, and the parallel port memory access handler. Place this into the appropriate MemAccess function, which for the ARM7TDMI model is MemAccessThumb. In the following example, the UART memory location corresponding to the UART_INIT_REGISTER address is being used to signal the initial IRQ (interrupt).

Access to the memory location opens uartin.txt for reading and uartout.txt for writing. The interrupt is scheduled for reading at IRQ_Time 40000 and for writing to UART_RCV_BUFFER after IRQ_Time 25000. The next section of the code identifies when the application code accesses a byte from UART_RCV_BUFFER or writes to UART_TXM_BUFFER.

Step 10: Rebuild the ARMulator, ensuring that models.o is rebuilt.

Example 6-1. Code for UART_PI_INTC.c

```
/*UART_PI_INTC.c - Fast ARMulator memory interface. */
#include <stdio.h>
#include <stdlib.h>
#include <string.h>   /* for memset */
#include <ctype.h>    /* for toupper */
#include "armdefs.h"
#include "armcnf.h"
#include "rdi.h"
#include "softprototype.h"
/* Step 3*/
#define ModelName (tag_t)"UART_PI_INTC"
 /*------------------------------------------*/
/* APB Peripherals */
#define  PI_INIT_REGISTER    0x200000
#define  PI_READ_REGISTER    0x200001
```

```
#define  UART_INIT_REGISTER  0x300000
#define  UART_RCV_BUFFER     0x300001
#define  UART_TXM_BUFFER     0x300005
 /*-------------------------------------------*/
#define   INTERRUPT_SOURCES  0x0a000200
/* Interrupt Setting */
/*--------------------------------------------*/
#define   INT_UNESED          1 << 0
#define   INT_PROGRAMMEDINT   1 << 1
#define   INT_DEBUG_RX        1 << 2
#define   INT_DEBUG_TX        1 << 3
#define   INT_TIMER1          1 << 4
#define   INT_TIMER2          1 << 5
#define   INT_PCCARD_A        1 << 6
#define   INT_PCCARD_B        1 << 7
#define   INT_SERIAL_A        1 << 8
#define   INT_SERIAL_B        1 << 9
#define   INT_PARALLEL        1 << 10
#define   INT_ASB0            1 << 11
#define   INT_ASB1            1 << 12
#define   INT_PI_APB0         1 << 13
#define   INT_UART_IN_APB1    1 << 14
#define   INT_UART_OUT_APB2   1 << 15

/* Step 2 */
typedef struct {
:
:
/* Added for PI */
int PI_IRQ;
FILE *pifile;

/* Added for UART */
int UART_IN_IRQ;
int UART_OUT_IRQ;
FILE *uart_infile;
FILE *uart_outfile;
:
:
} toplevel;

/* Step 5 */
/*---------------------------------------------*/
/* Callbacks for PI, UART interrupts */

extern unsigned Common_Irq(void* handle)
{
toplevel *top = (toplevel*) handle;
```

```
ARM_IRQ_Controller.nIRQStatus                          =
ARM_IRQ_Controller.nIRQRawStatus                       &
ARM_IRQ_Controller.nIRQEN ;
ARMul_ConsolePrint(top->state,"INTC  RAW  STATUS  :  %x
\n",ARM_IRQ_Controller.nIRQRawStatus );
ARMul_ConsolePrint(top->state,"INTC  STATUS  :  %x  \n",
ARM_IRQ_Controller.nIRQStatus );
ARMul_ConsolePrint(top->state,"INTC  ENABLE  :  %x  \n",
ARM_IRQ_Controller.nIRQEN );
ARMul_ConsolePrint(top->state,"INTC CLEAR      : %x \n",
ARM_IRQ_Controller.nIRQEC );

    if ( ARM_IRQ_Controller.nIRQStatus )
     {
      ARMul_ConsolePrint(top->state,"Set Interrupt \n");
      ARMul_SetNirq(top->state,LOW);
      return 1;
      }
     else
      {
    ARMul_ConsolePrint(top->state,"Clear Interrupt \n");
    ARMul_SetNirq(top->state,HIGH);
    return 0 ;
    }
}

extern unsigned pi_set_irq(void* handle)
{
 toplevel *top = (toplevel*) handle;
 ARM_IRQ_Controller.nIRQRawStatus |= INT_PI_APB0 ;
 ARMul_ConsolePrint(top->state,"AnIRQ has occured for PI
 :Raw Status : %x\n", ARM_IRQ_Controller.nIRQRawStatus);
  top->PI_IRQ = 1;
  Common_Irq(handle);
  return 1;
}
extern unsigned uart_set_in_irq(void* handle)
{
 toplevel *top = (toplevel*) handle;
 ARM_IRQ_Controller.nIRQRawStatus |= INT_UART_IN_APB1 ;
 ARMul_ConsolePrint(top->state,"AnIRQ has occured from
 Uart Receive \n");
 top->UART_IN_IRQ = 1;
 Common_Irq(handle);
 return 1;
}
extern unsigned uart_set_out_irq(void* handle)
{
  toplevel *top = (toplevel*) handle;
```

```
ARM_IRQ_Controller.nIRQRawStatus |= INT_UART_OUT_APB2 ;
ARMul_ConsolePrint(top->state,"AnIRQ has occured for
Uart Transmit\n");
   top->UART_OUT_IRQ = 1;
   Common_Irq(handle);
    return 1;
}

/* Step 6 */
/*-------------------------------------------------*/
static ARMul_Error MemInit(ARMul_State
*state,ARMul_MemInterface *interf, ARMul_MemType
type,toolconf config);
 ARMul_MemStub UART_INTC = {
  MemInit,
  ModelName
};
:
:
/* Step 7 */
static ARMul_Error MemInit(ARMul_State *state,
                           ARMul_MemInterface *interf,
                           ARMul_MemType type,
                           toolconf config)
{
:
:

  top->state=state;
  top->PI_IRQ=0;
  top->pifile=NULL;
  top->lddfile=NULL;
  top->UART_IN_IRQ=0;
  top->UART_OUT_IRQ=0;
  top->uart_infile=NULL;
  top->uart_outfile=NULL;
  ARM_IRQ_Controller.nIRQStatus = 0 ;
  ARM_IRQ_Controller.nIRQRawStatus = 0 ;
  ARM_IRQ_Controller.nIRQEN = 0 ;
  ARM_IRQ_Controller.nIRQEC = 0 ;
  ARM_IRQ_Controller.nIRQSoft = 0 ;
ARMul_ConsolePrint(top>state,"*****************\n");
ARMul_ConsolePrint(top->state,"***ARMulator with UART,
Timer and Parallel I/O port Models***\n");
  return ARMulErr_NoError;
}

:
:
```

```
/* Step 9 */
static int MemAccessThumb(void *handle,
                          ARMword address,
                          ARMword *data,
                          ARMul_acc acc)
{
  toplevel *top=(toplevel *)handle;
  unsigned int pageno;
   mempage *page;
  ARMword *ptr;
  ARMword offset;
  char PIfile[] = "C:\\tmp\\pi.txt";
  char UartIn[] = "C:\\tmp\\uartin.txt";
  char UartOut[] = "C:\\tmp\\uartout.txt";
  int IRQ_Time;
  /* INTERRUPT SOURCES register , Read access*/
  if((address == INTERRUPT_SOURCES) && (acc_WRITE(acc)))
  {
   ARMul_ConsolePrint(top->state, "Write to Interrupt
   Sources Register: %x\n", *data);
   ARM_IRQ_Controller.nIRQRawStatus = *data ;
   ARM_IRQ_Controller.nIRQStatus =
   ARM_IRQ_Controller.nIRQEN & (*data) ;
   Common_Irq(handle);
   return 1;
  }
 /* INTERRUPT SOURCES register , Read access*/

  if((address == INTERRUPT_SOURCES) && (acc_READ(acc)))
  {
     *data = ARM_IRQ_Controller.nIRQRawStatus ;
    ARMul_ConsolePrint(top->state, "Read from Interrupt
     Sources Register: %x\n", *data);
     return 1;
  }
/******Interrupt Controller Registers*****/

  /* IRQStatus register , read only */
  if((address == INTC_IRQ_STATUS) && (acc_READ(acc)))
  {
   ARMul_ConsolePrint(top->state, "Read from INTC Status
   Register \n");
   *data = (int)ARM_IRQ_Controller.nIRQStatus;
   return 1;
  }
  /* IRQRawStatus register , read only*/
  if((address == INTC_IRQ_RAWSTATUS) && (acc_READ(acc)))
  {
```

```
 ARMul_ConsolePrint(top->state, "Read from IRQRawSta-
 tus Register\n");
 *data = (int)ARM_IRQ_Controller.nIRQRawStatus;
 return 1;
}
/* IRQEnable register , read and write*/
if(address == INTC_IRQ_ENABLE)
{
        if(acc_READ(acc)) //read
        {
          ARMul_ConsolePrint(top->state, "Read
           IRQ_Enable Register\n");
          *data = (int)ARM_IRQ_Controller.nIRQEN;
        }
        else //write
        {
        ARMul_ConsolePrint(top->state, "Write
        IRQ_Enable Register\n");
        ARM_IRQ_Controller.nIRQEN = *data;
        ARM_IRQ_Controller.nIRQStatus =
        ARM_IRQ_Controller.nIRQRawStatus & (*data) ;
              Common_Irq(handle);
        }
              return 1;
}
/* IRQEnable register , read and write*/
if( (address == INTC_IRQ_CLEAR) && (acc_WRITE(acc)) )
{
 ARMul_ConsolePrint(top->state, "Write to IRQ ENABLE
 CLEAR Register\n");
 ARM_IRQ_Controller.nIRQEC = *data;
 ARM_IRQ_Controller.nIRQEN &=
~ARM_IRQ_Controller.nIRQEC ;
  ARM_IRQ_Controller.nIRQEC = 0x0 ;
 ARM_IRQ_Controller.nIRQStatus =
 ARM_IRQ_Controller.nIRQRaw
 Status & ARM_IRQ_Controller.nIRQEN ;
               Common_Irq(handle);
               return 1;
}
 /* IRQSoft register , Write only*/
if((address == INTC_IRQ_SOFT) && acc_WRITE(acc) )
{
    ARMul_ConsolePrint(top->state, "Write to IRQ_Soft
    Register\n");
    ARM_IRQ_Controller.nIRQSoft = *data;
    return 1;
}
```

```
/*******End of Interrupt Controller Registers******/
/********PI Memory Acess Handler******************/
if (( address == PI_INIT_REGISTER) && acc_READ(acc) )
   {
    top->pifile = NULL;
    ARMul_ConsolePrint(top->state,"PI : Trying to open:
    %s \n", PIfile);
    top->pifile = fopen(PIfile,"rb");
    if (top->pifile == NULL)
        {
         ARMul_ConsolePrint(top->state,"PI : Error:
         Could not open %s \n", PIfile);
         ARMul_ConsolePrint(top->state,"PI : No inter
         rupts will be scheduled.\n");
        }
else
        {
         ARMul_ConsolePrint(top->state,"PI : %s file
          successfully opened. \n", PIfile);
                 IRQ_Time = 10000 ;
         ARMul_ConsolePrint(top->state,"PI : An inter-
         rupt has been scheduled at %d \n",IRQ_Time);
         ARMul_ScheduleEvent(top->state,IRQ_Time,
          pi_set_irq, top);
        }
    *data = 1234; /* Fill in dummy return value*/
    return 1; /* indicating successful memory access*/
   }
//ARMul_ConsolePrint(top->state,"After  : Address  =  %x
Data = %x    Access = %d \n", address, *data, acc);
if ((address == PI_READ_REGISTER ) && (acc_READ(acc)))
   { ARMul_ConsolePrint(top->state,"PI : Read from
     PI_READ_REGISTER\n");
         if (top->PI_IRQ != 0)
             {
                 ARMul_SetNirq(top->state,HIGH);
                ARMul_ConsolePrint(top->state,"PI :
                 Clear Interrupt \n");
                 top->PI_IRQ = 0;
              }
                 if(top->pifile && !feof(top->pifile)) {
                        *data = fgetc(top->pifile);
         ARMul_ConsolePrint(top->state,"PI : Character
          read from PI file : %c \n",*data);
                        if ( !feof(top->pifile) )  {
                        IRQ_Time = 15000;
        ARMul_ConsolePrint(top->state,"PI : An interrupt
        has been scheduled at %d \n",IRQ_Time);
```

```
        ARMul_ScheduleEvent(top->state,IRQ_Time,
        pi_set_irq, top);
        ARMul_SetNirq(top->state,HIGH);
      // Not required if have an interrupt Controlle
      //ARMul_ConsolePrint(top->state,"PI : After Read :
      //Clear Interrupt \n");
      // top->PI_IRQ = 0;
                        }
                 }
        else
        ARMul_ConsolePrint(top->state,"PI : PI file is
        not open\n");
    return 1;
   }


/*******UART Memory Access Handler***********/
if (( address == UART_INIT_REGISTER) && (acc_WRITE(acc)
))
   {
    top->uart_infile = NULL;
    top->uart_outfile = NULL;
    ARMul_ConsolePrint(top->state,"Trying to open : %s
    and %s\n", UartIn, UartOut);
    top->uart_infile = fopen(UartIn,"rb");
    top->uart_outfile = fopen(UartOut,"wb");
    if (top->uart_infile == NULL)
       {
         ARMul_ConsolePrint(top->state,"UART : Error:
         Could not open %s \n", UartIn);
         ARMul_ConsolePrint(top->state,"UART : No inter-
         rupts will be scheduled.\n");
       }
    else if(top->uart_outfile == NULL)
       {
         ARMul_ConsolePrint(top->state,"UART : Error:
         Could not open %s \n", UartOut);
         ARMul_ConsolePrint(top->state,"UART : No inter-
         rupts will be scheduled.\n");
       }
   else
       {
        ARMul_ConsolePrint(top->state,"UART : %s and %s
        files successfully opened.\n", UartIn,UartOut);
        IRQ_Time = 40000;
        ARMul_ConsolePrint(top->state,"UART:An inter-
       rupt has been scheduled at time %d \n",IRQ_Time);
        ARMul_ScheduleEvent(top->state, IRQ_Time,
        uart_set_in_irq, top);
```

```
        IRQ_Time = 25000;
       ARMul_ConsolePrint(top->state,"UART:An interrupt
        has been scheduled at time %d \n", IRQ_Time);
        ARMul_ScheduleEvent(top->state,IRQ_Time,
         uart_set_out_irq, top);
       }
    *data = 1234; /* Fill in dummy return value*/
    return 1; /* indicating successful memory access*/
   }

//ARMul_ConsolePrint(top->state,"After  :  Address  =  %x
//Data = %x    Access = %d \n", address, *data, acc);

if (( address == UART_RCV_BUFFER) && (acc_READ(acc) ))
   {
        ARMul_ConsolePrint(top->state,"UART : Read from
         UART_RCV_BUFFER\n");
         if (top->UART_IN_IRQ != 0)
             {
                 ARMul_SetNirq(top->state,HIGH);
                 top->UART_IN_IRQ = 0;
              }
       if(top->uart_infile && !feof(top->uart_infile))
              {
                 *data = fgetc(top->uart_infile);
                 ARMul_SetNirq(top->state,HIGH);
                //Not required if we have an
                 //interrupt Controller
                 if (!feof(top->uart_infile) ) {
                 ARMul_ScheduleEvent(top->state,
                 NO_OF_CYCLES,uart_set_in_irq, top);
                        }
               }
        else
        ARMul_ConsolePrint(top->state,"UART : Codec in
        file is not open\n");
    return 1;
   }
if (( address == UART_TXM_BUFFER) && (acc_WRITE(acc) ))
   {
       ARMul_ConsolePrint(top->state,"UART : Write
       to UART_TXM_BUFFER\n");
       if (top->UART_OUT_IRQ != 0)
           {
               ARMul_SetNirq(top->state,HIGH);
               top->UART_OUT_IRQ = 0;
           }
       ARMul_ScheduleEvent(top->state, NO_OF_CYCLES,
       uart_set_out_irq, top);
```

```
        if (top->uart_outfile)
          fputc(*data, top->uart_outfile);
        else
          ARMul_ConsolePrint(top->state,"UART : Codec
            out file is not open\n");
        return 1;
    }
/*-----------------------------------------------*/
```

6.3.3.2 Writing the Application Code

The application code is written for data transfers between the UART and the processor, and it contains an IRQ (interrupt) handler for the parallel port, UART receive interrupt, and UART transmit interrupt. For each IRQ handler, the interrupt source is first cleared, then higher or equal priority interrupts are enabled. After the interrupt is serviced, the previous interrupt is restored. For the parallel port (PI), contents of `Pi_Read_Address` memory location are read and placed in a global variable. For UART in interrupt, the contents of `UART_RCV_BUFFER` are read and placed in the variable `Sample` for further processing. For UART out interrupt, contents of the processed value from `Sample` are stored in `UART_TXM_BUFFER`.

The `Install Handler` code installs a branch instruction in the IRQ exception vector to branch to the IRQ handler. The main function installs the IRQ handler and then accesses the memory location for the PI and UART to initiate the interrupt. The program then goes into a loop until the first interrupt occurs, at which point the program flow diverts to the IRQ handler.

The application can be run on the host system. The debugger running on the host can be used to single-step or set breakpoints in the software to analyze the critical part of the software for intended functionality.

Example 6-2. Code for Data Transfer between the UART and Processor

```
#include <stdio.h>

/*--------------------------------------------*/
/* APB Peripherals */
/*--------------------------------------------*/
#define UART_IN_INT                  1
#define UART_OUT_INT                 2
#define PI_INT                       3
```

```
#define PI_INIT_REGISTER        0x200000
#define PI_READ_REGISTER        0x200001
#define UART_INIT_REGISTER      0x300000
#define UART_RCV_BUFFER         0x300001
#define UART_TXM_BUFFER         0x300005
/*--------------------------------------------------*/
unsigned *Pi_Init_Address =(unsigned*)PI_INIT_REGISTER;
unsigned *Pi_Read_Address =(unsigned*)PI_READ_REGISTER;
unsigned *Uart_Init_Address=
                        (unsigned*)UART_INIT_REGISTER;
unsigned *Rcv_Buf_Address =(unsigned*)UART_RCV_BUFFER;
unsigned *Txm_Buf_Address =(unsigned*)UART_TXM_BUFFER;
/*---------------------------------------------------*/
/* ARM Interrupt Controller Register memory map */
/*---------------------------------------------------*/
#define INTC_BASE_ADD      0x0a000000
#define INTC_IRQ_STATUS    0x0a000000  /*Read*/
#define INTC_IRQ_RAWSTATUS 0x0a000004  /*Read*/
#define INTC_IRQ_ENABLE    0x0a000008  /*Read/Write*/
#define INTC_IRQ_CLEAR     0x0a00000c  /*Write*/
#define INTC_IRQ_SOFT      0x0a000010  /*Write*/
#define INTC_FIQ_STATUS    0x0a000100  /*Read*/
#define INTC_FIQ_RAWSTATUS 0x0a000104  /*Read*/
#define INTC_FIQ_ENABLE    0x0a000108  /*Read/Write*/
#define INTC_FIQ_CLEAR     0x0a00010c  /*Write*/
#define INTC_FIQ_SOURCE    0x0a000114  /*Read/Write*/
/*---------------------------------------------------*/
#define INTERRUPT_SOURCES  0x0a000200
/*---------------------------------------------------*/
/*   Interrupt Seting  */
/*---------------------------------------------------*/
#define INT_UNESED  1 << 0
#define INT_PROGRAMMEDINT  1 << 1
#define INT_DEBUG_RX       1 << 2
#define INT_DEBUG_TX       1 << 3
#define INT_TIMER1         1 << 4
#define INT_TIMER2         1 << 5
#define INT_PCCARD_A       1 << 6
#define INT_PCCARD_B       1 << 7
#define INT_SERIAL_A       1 << 8
#define INT_SERIAL_B       1 << 9
#define INT_PARALLEL       1 << 10
#define INT_ASB0           1 << 11
#define INT_ASB1           1 << 12
#define INT_PI_APB0        1 << 13
#define INT_UART_IN_APB1   1 << 14
#define INT_UART_OUT_APB2  1 << 15

/*------------------------------------------------------*/
```

```
/*  Mask For Priority Seting  */
/*---------------------------------------------------*/
#define MASK_UNUSED             INT_UNESED
#define MASK_PROGRAMMEDINT MASK_UNUSED +
                            INT_PROGRAMMEDINT
#define MASK_DEBUG_RX   MASK_PROGRAMMEDINT +
                        INT_DEBUG_RX
#define MASK_DEBUG_TX   MASK_DEBUG_RX + INT_DEBUG_TX
#define MASK_TIMER1     MASK_DEBUG_TX + INT_TIMER1
#define MASK_TIMER2     MASK_TIMER1 + INT_TIMER2
#define MASK_PCCARD_A   MASK_TIMER2 + INT_PCCARD_A
#define MASK_PCCARD_B   MASK_PCCARD_A + INT_PCCARD_B
#define MASK_SERIAL_A   MASK_PCCARD_B + INT_SERIAL_A
#define MASK_SERIAL_B   MASK_SERIAL_A + INT_SERIAL_B
#define MASK_PARALLEL   MASK_SERIAL_B + INT_PARALLEL
#define MASK_ASB0       MASK_PARALLEL + INT_ASB0
#define MASK_ASB1       MASK_ASB0 + INT_ASB1
#define MASK_PI_APB0    MASK_ASB1  + INT_PI_APB0
#define MASK_UART_IN_APB1  MASK_PI_APB0 +
                           INT_UART_IN_APB1
#define MASK_UART_OUT_APB2  MASK_UART_IN_APB1  +
                            INT_UART_OUT_APB2
/*-------------------------------------------------*/

volatile int GlobalVar ;
int Sample =0 ;
void __irq myIRQhandler(void)
{
 int nType, Enable ;
 char *IRQType = (char*)INTC_IRQ_STATUS;
 int *address ;
 nType = *IRQType;
 if ( nType & INT_PI_APB0)/* PI Interrupt Detected */
   {
     address  = (int*)INTERRUPT_SOURCES ;
     *address  ^= INT_PI_APB0 ;/* Clear PI Interrupt
     Source */
     address  = (int*)INTC_IRQ_ENABLE ;/* Enable higher
     or equal priorities */
     Enable   = *address ;
     *address = MASK_PI_APB0 ;
     address  = (int*)INTC_IRQ_CLEAR ;
     *address  = INT_PI_APB0 ; /* Clear PI Interrupt */
     /* body of the ISR + Clear Interrupt Source */
     GlobalVar = *((char*)Pi_Read_Address);
     address = (int*)INTC_IRQ_ENABLE ;/* Restore Previ-
     ous Interrupt */
     *address = Enable;
     return ;
```

```
        }
  if ( nType & INT_UART_IN_APB1) /* UART_in Interrupt
  Detected */
    {
      address  = (int*)INTERRUPT_SOURCES ;
      *address  ^= INT_UART_IN_APB1 ; /* Clear UART_in
      Interrupt Source */
      address  = (int*)INTC_IRQ_ENABLE ;/* Enable higher
      or equal priorities */
      Enable   = *address ;
      *address = MASK_UART_IN_APB1 ;
      address  = (int*)INTC_IRQ_CLEAR ;
      *address  = INT_UART_IN_APB1 ; /* Clear UART_in
      Interrupt */
      /* body of the ISR + Clear Interrupt Source */
      GlobalVar = *((char*)Rcv_Buf_Address);
      Sample = GlobalVar ;
      address = (int*)INTC_IRQ_ENABLE ;/* Restore Previ-
      ous Interrupt */
      *address = Enable;
      return ;
     }

  if ( nType & INT_UART_OUT_APB2)  /* UART_out Interrupt
  Detected */
    {
      address  = (int*)INTERRUPT_SOURCES ;
      *address  ^= INT_UART_OUT_APB2 ;/* Clear UART_out
      Interrupt Source */
      address  = (int*)INTC_IRQ_ENABLE ;/* Enable higher
      or equal priorities */
      Enable   = *address ;
      *address = MASK_UART_OUT_APB2 ;
      address  = (int*)INTC_IRQ_CLEAR ;
      *address  = INT_UART_OUT_APB2 ; /* Clear UART_out
      Interrupt */
      GlobalVar = Sample;    /* body of the ISR + Clear
      Interrupt Source */
      *Txm_Buf_Address  = Sample ;
      address = (int*)INTC_IRQ_ENABLE ; /* Restore Pre-
      vious Interrupt */
      *address = Enable;
      return ;
     }
 return ;
 }

unsigned  Install_Handler(unsigned  routine,  unsigned
*vector)
```

```
{
unsigned vec, oldvec ;
vec = ((routine - (unsigned)vector -0x8) >> 2);
vec = 0xea000000 | vec ;  /* Build up Branch Always ARM
instruction for exception vector*/
oldvec = *vector;   /*Read  current value from vector*/
*vector = vec;      /*install new vector*/
return (oldvec);    /*return the old vector*/
}

int main()
{
int value ;
int *address ;

/* Define the location of IRQ vector*/
unsigned *IrqVec = (unsigned *)0x18;

/* Install interrupt handler for PI, UART In and UART
out*/
Install_Handler((unsigned)myIRQhandler, IrqVec);

/*Access the memory location whichwill initialize the
interrupt*/
value = *Pi_Init_Address ;   /* init PI by reading from
Init Register*/
*Uart_Init_Address = 0x45 ;  /* init UART  by writting
to Uart init register*/

address  = (int*)INTC_IRQ_ENABLE ;/* Enable higher or
equal priorities */
*address  =  INT_UART_IN_APB1  |  INT_PI_APB0  |
INT_UART_OUT_APB2 ;

printf("Enable : %x \n",*address);

/* the do while loop will continue until a full stop
char from the text file*/
  do {
        GlobalVar = -1;
        while (GlobalVar == -1);
        printf("Character read : %c\n", GlobalVar);
   }
  while (GlobalVar != '.');
  printf("END OF PROGRAM");
}
```

6.4 Co-verification

Co-verification provides the ability to integrate and verify hardware and software concurrently. Figure 6-7 shows a simple block diagram of co-verification environment.The processor is replaced by an ISS that interfaces with the source-level debugger and provides an interface to the peripherals through a processor BFM. The peripherals are modelled with RTL code in Verilog or VHDL.

The objectives of co-verification are:

- To eliminate design errors before the silicon is fabricated
- To enable engineers to design, develop, and debug both hardware and software simultaneously
- To achieve fast turn-around time in the design cycle

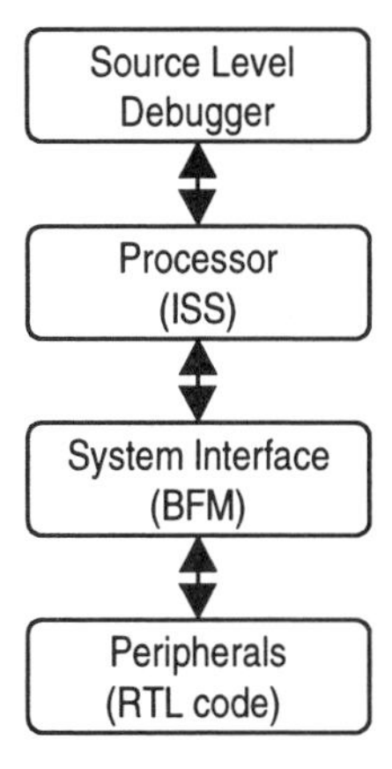

Figure 6-7. Co-verification Environment

The co-verification environment provides the following features for debugging:

- Displays the software source code and processor state
- Configures the processor memory and I/O address map, enabling accessibility to hardware components
- Synchronizes software and hardware clock cycles
- Detects the interrupts generated by the peripherals using the ISS model

- ISS can perform read/write on the memory instantiated in the hardware
- Facility to set breakpoints or single-stepping in both software and hardware description language (HDL) code for detailed analysis and debug

6.4.1 Co-verification Environment

Figure 6-8 shows the basic components of the co-verification environment.

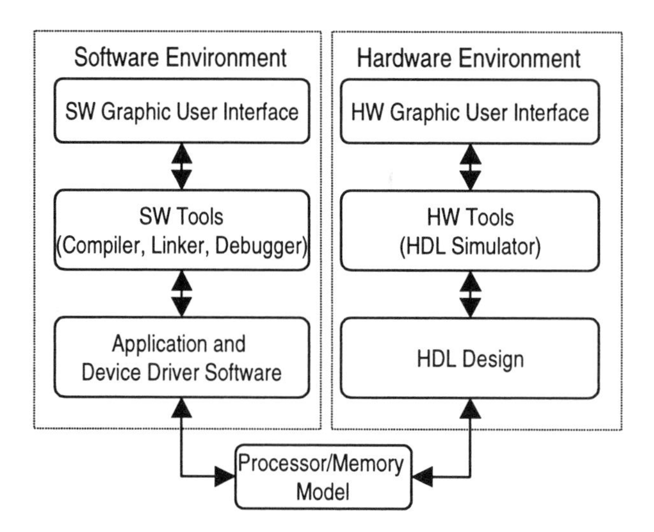

Figure 6-8. Co-verification Environment

The software environment consists of a graphical user interface (GUI) and software development tools, such as compilers, linkers, and debuggers. The application and device driver software are loaded into the host system memory. The co-verification environment does not put any restrictions and also does not call for any special software development techniques to be followed. The software developed and tested using co-verification can be used to test the final prototype system with no modifications.

The hardware environment consists of a GUI, hardware simulator, and HDL design. The software and hardware interactions take place through a processor/ memory model. The processor model comprises of an ISS, and a BFM. The ISS is a software module that models the functional behavior of a processor's instruction set. The BFM simulates the activities at the periphery of the processor in the simulation environment. The software calls are translated as test vectors to stimulate the hardware simulator. The hardware simulator responds and outputs the results back to the software environment. The outputs can be observed using the waveform viewer embedded within the hardware simulator.

It is recommended to begin simulating the basic functions in the hardware simulator. As the confidence level is increased and the basic initialization software is developed, the design can be brought into the co-verification environment for HW/SW integrated verification. If errors are found while running the software on a hard prototype or other verification environments (such as emulation or a hardware accelerator), the failing software can be run in the co-verification environment for detailed analysis and debugging.

6.4.2 Selecting a Co-verification Environment

The following considerations should be taken into account when selecting a co-verification environment.

- **Performance**: The performance of the simulation is very crucial for the design, because the tools need to handle a great amount of software and hardware data, depending on the application. Tool vendors claim performance in the range of 200 to 500 instructions per second (IPS).
- **Availability of the processor models**: Should support the processors being incorporated within the SOC design. Today's co-verification tools support many of the popular embedded core processors, such as ARM, MIPS, PowerPC, DSP, etc.
- **Functions of the processor model**: Some of the processor models do not support the full suite of functions that a specific core can provide (for example, pipelining may not be supported by all models). Check that the functions of the processor required for the application are supported by the model supported by the co-verification tool.
- **Accuracy and completeness of processor models**: The performance of the co-verification depends on the mode used. Two modes of accuracy are available:

  Functional accuracy—This involves running the software on a hardware model. No timing issues are addressed. Sufficient for debugging and testing the software at source-code level.

  Cycle accuracy—The software is cross-compiled to generate a executable on the target system and run on a cycle-accurate ISS against the hardware model.

The ISS can keep track of the number of simulation clock cycles for hardware transactions. The timing of critical sections of code can be verified.

Functional accuracy produces higher performance than cycle accuracy.

- **RTOS support**: Should support the RTOS required for the application. For example, pSOS and VxWorks are supported by many of the commercial co-verification environments.
- **Multiple processor support**: Should support multiple processor cores in a single design if multiple processor cores need to be embedded within the same chip.

6.4.2.1 Limitations of Co-verification Tools

The limitations of co-verification tools available today are:

- **Availability of accurate processor models**: Some of the processor models do not provide full functionality of the processor, such as pipelining functions in the processor.
- **Performance**: Sufficient performance to run the complete application software on top of the target RTOS is not available because of capacity and simulation speed limitations. These tools do offer sufficient performance to run the interface confidence tests, code segments, and individual driver and utility code.
- **Support of software functions**: Some tools do not support the array-based functions that may be used in application software code.

6.4.3 Co-verification Methodology

Figure 6-9 shows the co-verification methodology flow for an ICE-based co-verification environment. The methodology involves the following steps.

1. **Processor model setup**: Required for debugging the design and is called by instantiating the I/O ports. The setup requires the processor identification number, debugger configuration file name, and debug memory configuration.
2. **Software model setup**: Involves describing the source code, generating the debug information files, and creating an executable image with the debug inormation. The executable image is created using the compiler, assembler, and linker.
3. **Configuration file setup**: Required for initial debugger setup. It involves specifying the file to be downloaded to the debug memory, and the script file for the debugger to start.

4. **Run simulation**: The software debugger and hardware simulator are run. The software debugger enables the user to set break-points in the software code and observe the contents in the processor registers and debug memory. The hardware signals can be probed using the hardware simulator. The contents of the specific register or memory location can also be viewed or set.

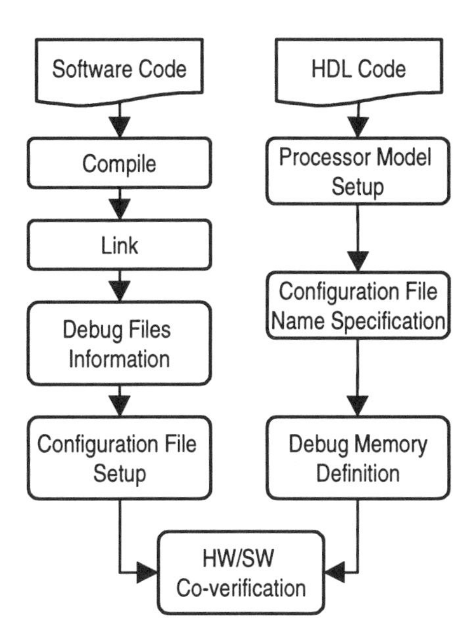

Figure 6-9. Co-verification Methodology Flow

6.4.4 UART Co-verification

Figure 6-10 shows a simple block diagram of the example Bluetooth SOC design. This design was verified using the co-verification environment. The highlighted hardware blocks are used to illustrate the example with the following software functions:

- Main routine running on ARM7TDMI processor
- Interrupt handler routine running on ARM7TDMI processor
- Device driver routine for UART

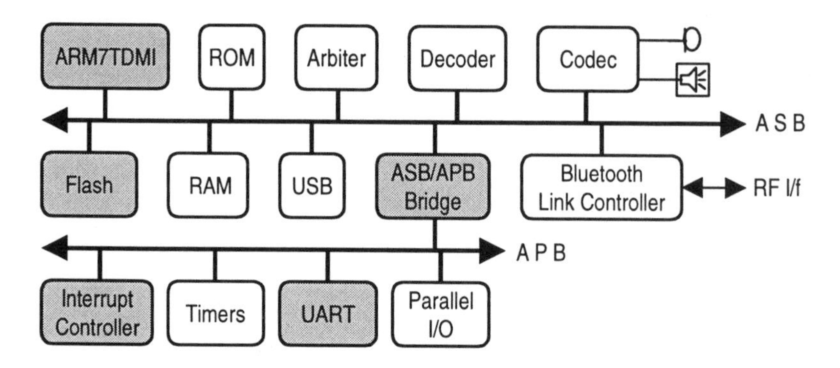

Figure 6-10. Block Diagram of Bluetooth SOC

Except for the UART block, the Bluetooth SOC design is developed in Verilog. The UART block is developed in VHDL and is used in the design by wrapping it with a Verilog shell.

6.4.4.1 UART Design Description

The UART described here is compatible with the 16C550 industry-standard specification. The UART interfaces the processor with an external device (peripheral) that transmits and receives the data in serial form. It consists of a transmitter, a receiver, a baud rate generator, system bus interface logic, interrupt logic, and transmit/receive control logic, as shown in Figure 6-11.

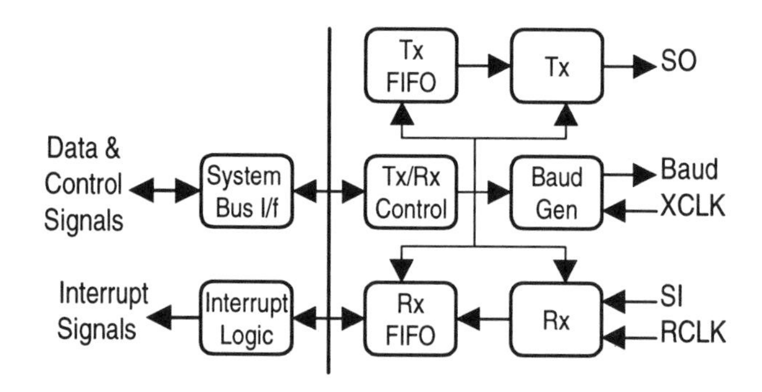

Figure 6-11. Block Diagram of UART

6.4.4.2 Transmitter

The transmitter converts the parallel data from the processor into a serial bit stream on the serial output (SO) pin. It consists of a TxFIFO and Tx buffer. It sends a start bit, a programmed number of data bits, a parity bit, and stop bits. The least significant bit is sent first. If a new character is not available in the TxFIFO after the transmission of the stop bits, the SO output remains at logic 1 (high), and the TxRDY bit in the status register will be set to 1. The TxRDY bit is cleared when a new character is loaded into the TxFIFO by the processor. If the transmitter is disabled, it continues operating until the character presently being transmitted and any characters in the TxFIFO, including the parity and stop bits have been transmitted. A reset assertion stops all transmission immediately, clearing the TxFIFO, status, and transmit interrupts. A software command can be used to reset the transmitter.

6.4.4.3 Receiver

The receiver converts the incoming serial bit stream into parallel data that is acceptable by the processor. The UART is programmed to receive data when enabled through the command register. The receiver looks for a start bit on the serial input (SI) pin. If a high-to-low transition is detected, the state of the SI pin is sampled each 16Xclock for 7-1/2 clocks (16X clock mode) or at the next rising edge of the bit time clock (1X clock mode). The receiver continues to sample the input at one-bit time intervals until all the data bits, parity bit, and stop bit have been received. The least significant bit is received first. The data is then transferred to the RxFIFO and the RxRDY bit in the status register is set. This condition can be programmed to generate an interrupt to the processor, indicating the availability of data from the peripheral connected to the UART.

The baud-rate generator is a free-running counter that generates x16 clocks. It provides timing information for UART transmit/receive control.

The interrupt logic provides a single interrupt to the system-interrupt controller. It ORs the interrupts generated by the transmit and receive blocks in the UART.

6.4.4.4 UART Register Definitions

The UART configuration registers are as follows:

- Receive Buffer Register (RBR), address: 0xA0600000—Read-only register that contains the next byte to be read by the processor. The data is received from the peripheral connected to the receiver.

- Transmit Hold Register (THR), address: 0xA0600000—Write-only register that holds a byte of data written from processor.
- Interrupt Enable Register (IER), address: 0xA0600001—Read/write register used to enable/disable the interrupt. The significance of each bit in the IER is:
  bit 0—RxFIFO interrupt

  bit 1—TxFIFO Interrupt

  bit 2—Rx line status,

  bit 3—Modem status

  bits 7-4 are reserved and return "0" when read by the processor
- Interrupt Status Register (ISR), address: 0xA0600002—Read-only register, providing interrupt status of the transmit and receive blocks.
- FIFO Control register (FCR), address: 0xA0600002—Write-only register that determines whether the TxFIFO and RxFIFO are enabled and in which mode in which line status interrupts are generated to the processor.
- Line Control Register (LCR), address: 0xA0600003—Read/write register. The seven-least significant bits of this register determine the characteristics of the transmitted and received serial data. For example, word length, number of stop bits, and parity encoding.
- Line Status Register (LSR), address: 0xA0600005—Read-only register, indicating the status of the receiver or transmitter. For example, if bit 0 is set, it indicates that data from the RxFIFO is available to be read by the processor.
- UART Divisor Latch (UDL), address: 0xA0600000—Read and write register that controls the bit transmission rate for serial output data.

6.4.4.5 UART RTL Module Port

Example 6-3 shows the RTL module port declaration for the UART that is developed in VHDL. The instantiation of the various blocks within the UART is also shown. However, the detail RTL code for the UART design is not included here.

Example 6-3. RTL Module Port for the UART

```
----     INPUTS :
--                   notresetR
--                   sysclk
--                   cs_n
--                   add
```

```
--                  re_n
--                  we_n
--                  di
--                  xclk
--                  do
--                  rclk
--                  si
--                  cts_n
--                  dsr_n
--                  ri_n
--                  dcd_n
--
--      OUTPUTS :   baudout
--                  so
--                  int
--                  txrdy_n
--                  rxrdy_n
--                  dtr_n
--                  rts_n
--                  op1_n
--                  op2_n
--
--
--   Top level
--
-- This block instantiates the blocks that are driven

-- by separate clocks, namely the BaudGen, Tx, Rx and
-- UART_sysclk blocks
----------------------------------------------------

library IEEE;
use IEEE.std_logic_1164.all;

entity UART is

    port(
         resetR_n   : in  std_logic;
         sysclk     : in  std_logic;
         cs_n       : in  std_logic;
         add        : in  std_logic_vector (2 downto 0);
```

```
          re_n        : in  std_logic;
          we_n        : in  std_logic;
          di          : in  std_logic_vector (7 downto 0);
          xclk        : in  std_logic;
          rclk        : in  std_logic;
          si          : in  std_logic;
          cts_n       : in  std_logic;
          dsr_n       : in  std_logic;
          ri_n        : in  std_logic;
          dcd_n       : in  std_logic
          int         : out std_logic;
          txrdy_n     : out std_logic;
          rxrdy_n     : out std_logic;
          dtr_n       : out std_logic;
          rts_n       : out std_logic;
          op1_n       : out std_logic;
           op2_n       : out std_logic;
          o_baudout : out std_logic;
          o_so         : out std_logic;
          o_do        : out std_logic_vector (7 downto 0)
         );
end UART;

architecture RTL of UART is

component BaudGen
    port(
          xclk    : in  std_logic;
          dl      : in  std_logic_vector (15 downto 0);
          reset   : in  std_logic;
          baudout: out std_logic
         );
end component;

component Tx
    port(
          reset     : in  std_logic;
          tclk      : in  std_logic;
          data      : in  std_logic_vector (7 downto 0);
          dav       : in  std_logic;
          wlength   : in  std_logic_vector (1 downto 0);
```

```
            stpbits   : in  std_logic;
            ptymode   : in  std_logic_vector (2 downto 0);
            txbreak   : in  std_logic;
            tsr_empty: out std_logic;
            sout      : out std_logic;
            ack       : out std_logic
          );
end component;

component Rx
   port(
           reset      : in  std_logic;
           rclk       : in  std_logic;
           sin        : in  std_logic;
           wlength    : in  std_logic_vector (1 downto 0);
           stpbits    : in  std_logic;
           ptyen      : in  std_logic;
           data       : out std_logic_vector (10 downto 0);
           dav        : out std_logic;
           time_char : out std_logic
         );
end component;

component UART_sysclk
 port(
        reset     : in  std_logic;
        sysclk    : in  std_logic;
        cs_n      : in  std_logic;
        add       : in  std_logic_vector (2 downto 0);
        re_n      : in  std_logic;
        we_n      : in  std_logic;
        di        : in  std_logic_vector (7 downto 0);
        tsr_empty: in std_logic;
        txack     : in std_logic;
        time_char: in std_logic;
        rxdata    : in std_logic_vector (10 downto 0);
        rxdav     : in std_logic;
        cts_n     : in  std_logic;
        dsr_n     : in  std_logic;
        ri_n      : in  std_logic;
        dcd_n     : in  std_logic
```

```
        do        : out std_logic_vector (7 downto 0);
        int       : out std_logic;
        rxrdy     : out std_logic;
        reg_dl    : out std_logic_vector (15 downto 0);
        txdata    : out std_logic_vector (7 downto 0);
        txdav     : out std_logic;
        reg_lcr   : out std_logic_vector (7 downto 0);
        loopback  : out std_logic;
        txfull    : out std_logic;
        dtr_n     : out std_logic;
        rts_n     : out std_logic;
        op1_n     : out std_logic;
        op2_n     : out std_logic
       );
end component;

-- Internal block instantiation


begin
  i_BaudGen: BaudGen
     port map(
                xclk    => xclk,
                dl      => reg_dl,
                reset   => resetR_n,
                baudout => txclk
            );

  i_Tx: Tx
     port map(
                reset      => resetR_n,
                tclk       => txclk,
                data       => txdata,
                dav        => txdav,
                wlength    => reg_lcr(1 downto 0),
                stpbits    => reg_lcr(2),
                ptymode    => reg_lcr(5 downto 3),
                tsr_empty  => tsr_empty,
                sout       => txout,
                ack        => txack
              );
```

```
i_Rx: Rx
   port map(
            reset       => resetR_n,
            rclk        => rxclk,
            sin         => rxin,
            wlength     => reg_lcr(1 downto 0),
            stpbits     => reg_lcr(2),
            ptyen       => reg_lcr(3),
            time_char   => time_char,
            data        => rxdata,
            dav         => rxdav
           );

i_sysclk_block: UART_sysclk
   port map(
            reset        => resetR_n,
            sysclk       => sysclk,
            cs_n         => cs_n,
            add          => add,
            re_n         => re_n,
            we_n         => we_n,
            di           => di,
            do           => do,
            tsr_empty    => tsr_empty,
            txack        => txack,
            time_char    => time_char,
            rxdata       => rxdata,
            rxdav        => rxdav,
            int          => int,
            rxrdy        => rxrdy_n,
            k            => reg_lcr(6),
            reg_dl       => reg_dl,
            otxdata       => txdata,
            txdav        => txdav,
            reg_lcr      => reg_lcr,
            loopback     => loopback,
            txfull       => txfull,
            dtr_n        => dtr_n,
            rts_n        => rts_n,
            op1_n        => op1_n,
```

```
                    op2_n          => op2_n,
                    cts_n          => cts_n,
                    dsr_n          => dsr_n,
                    ri_n           => ri_n,
                    dcd_n          => dcd_n
                    );

-- Top-level signal assignments
    rxin   <= txout when (loopback = '1')
         else si;

    rxclk <= txclk when (loopback = '1')
           else rclk;


-- Output assignments
    so       <= '1' when (loopback = '1')
                else txout;
    baudout <= txclk;
    txrdy_n <= txfull;
    resetR_n <= not notresetR;
end RTL;
```

6.4.4.6 Verilog Shell for UART VHDL Design

Example 6-4 shows the shell file for the UART VHDL model that is used for simulation with the Verilog simulator. For example, in the Cadence NC-Verilog simulator, the ncshell utility converts the VHDL interface port types to Verilog data types, and a Verilog shell is generated for the VHDL model. This enables the VHDL model to be imported to Verilog for simulation.

Example 6-4. Verilog Shell File for UART VHDL Model

```
module uart(
resetR_n,
sysclk,
cs_n,
add,
```

```
re_n,
we_n,
di,
do,
xclk,
baudout,
so,
rclk,
si,
int,
txrdy_n,
rxrdy_n,
dtr_n,
rts_n,
op1_n,
op2_n,
cts_n,
dsr_n,
ri_n,
dcd_n
)

(*    const    integer    foreign    =    "VHDL(event)
WORKLIB.UART:rtl"; *);

input  resetR_n;
input  sysclk;
input  cs_l;
input  [2:0] add;
input  re_n;
input  we_n;
input  [7:0] di;
output [7:0] do;
input  xclk;
output baudout;
output so;
input  rclk;
input  si;
output int;
output txrdy_n;
output rxrdy_n;
```

```
output dtr_n;
output rts_n;
output op1_n;
output op2_n;
input  cts_n;
input  dsr_n;
input  ri_n;
input  dcd_n;

endmodule
```

6.4.4.7 Hardware Testbench for the UART

To simulate the incoming data to the receiver part of the UART, a suitable data stream using a testbench needs to be generated. The code in Example 6-5 gives excerpts of the stimulus applied to the UART from the top-level testbench for the Bluetooth SOC design.

Example 6-5. Excerpts of UART Hardware Testbench

```
reg mem_uart[16383:0];  // Address for UART input file
initial
begin
  $readmemh ( "./TEST_FILE.txt", mem );
  $readmemh ( "./UART.txt",mem_uart);
  Top.Bluetooth.uart_si = 1;
// initialize UART serial input to 0.
  Top.Bluetooth.uart_rclk = 0;
// initialize UART receive clock to 0.
  Top.Bluetooth.uart_xclk = 0;
//initialize UART transmit clock to 0.
end

parameter uart_clk_period = 542;
// Clock period for Maximum baudrate of 115200
// 1.843 MHZ UART frequency
parameter UART_CLK_DIV = 2;
```

```
always
begin
  #(uart_clk_period/2);
  Top.Bluetooth.uart_rclk = ~Top.Bluetooth.uart_rclk;
  Top.Bluetooth.uart_xclk = ~Top.Bluetooth.uart_i_xclk ;
end

/******************UART Input******************/
initial
begin
  #250000;
  cnt = 14'h0;
  while (cnt <= 5)
  begin
  Bluetooth.uart_si = 0;
  #(uart_clk_period*UART_CLK_DIV);
  cnt = cnt +1;
  end
  cnt = 0;
  while ( cnt <= 164 )
  begin
    $display ( "Sent %h to   UART at %0t", mem_uart[cnt]
    , $time );
    Bluetooth.uart_si =  mem_uart[cnt] ;
    #(16*uart_clk_period*UART_CLK_DIV);
    cnt = cnt + 1;
  end
 end
```

6.4.4.8 Software Code for Testing the UART

The software code for testing the UART consists of header files and the following main functions:

- Main routine running on the ARM7TDMI processor
- Interrupt handler routine running on the ARM7TDMI processor
- Device driver routine for the UART

6.4.4.9 Software Header Files

The software requires the following header files. The header files define the various parameters used in the software. Example 6-6 shows the Bluetooth SOC design parameters, the peripheral device parameters, and the peripheral interrupt parameters used.

Example 6-6. Bluetooth SOC Parameters

Bluetooth.h

```
#include "interrupt.h"
#define  TRUE                 1
#define  FALSE                0
#define outl(Port, Value) (*((volatile long *)(Port))=
                            Value)
#define inl(Port)            (*((volatile long *)(Port)))
#define outc(Port, Value) (*((volatile char *)(Port))=
                            Value)
#define inc(Port)            (*((volatile char *)(Port)))
#define outs(Port, Value) (*((volatile short int*)
                           (Port))=Value)
#define  ins(Port)          (*((volatile short int *)
                            (Port)))
#define UART_IO_SIZE          4
#define BUF_SIZE              4
```

Peripheral device parameters

Devices.h

```
#define TMR_REG                 0xa0500000
#define CODEC_REG               0xa0700000
#define PIO_REG                 0xa0600000
#define UART_BASE               0xa1c00000
#define UART_RCV_BFR_REG        UART_BASE
#define UART_INT_ENABLE_REG     UART_BASE + 0x01
#define UART_FIFO_CNTRL_REG     UART_BASE + 0x02
#define UART_INTR_ST_REG        UART_BASE + 0x02
#define UART_LINE_ST_REG        UART_BASE + 0x05
#define UART_LINE_CNTRL_REG     UART_BASE + 0x03
```

```
#define UART_MCR                 UART_BASE + 0x04
#define UART_DLL_REG             UART_BASE
#define UART_CLK_DIV             0x0202
```

Peripheral interrupt parameters

Interrupt.h

```
#define INT_CTRL_BASE               0xA0100000
#define IRQ_STATUS                  INT_CTRL_BASE
#define IRQ_RAW_STATUS              (INT_CTRL_BASE + 0x04)
#define IRQ_ENABLE                  (INT_CTRL_BASE + 0x08)
#define IRQ_ENABLE_SET              (INT_CTRL_BASE + 0x08)
#define IRQ_ENABLE_CLR              (INT_CTRL_BASE + 0x0c)
#define IRQ_SOFT                    (INT_CTRL_BASE + 0x10)
#define PRIORITY_0                  0
#define PRIORITY_1                  1
#define PRIORITY_2                  2
#define PRIORITY_3                  3
#define PRIORITY_UART               4
#define PRIORITY_TMR                6
#define PRIORITY_PI                 7
#define IRQ_UART_BIT                1<<PRIORITY_UART
#define IRQ_TMR_BIT                 1<<PRIORITY_TMR
#define IRQ_PI_BIT                  1<<PRIORITY_PI
#define  MASK_PRIORITY_4                  MASK_PRIORITY_5   |
IRQ_UART_BIT
#define MASK_PRIORITY_6            MASK_PRIORITY_7  |
                                    IRQ_TMR_BIT
#define MASK_PRIORITY_7             IRQ_PI_BIT
```

6.4.4.10 Software Routines for the UART Test

The software routines used in the UART test are described below.

Main ARM Routine

This routine, which runs on the ARM7TDMI processor, configures the UART, initializes the transmit/receive buffers, and waits for an interrupt from the UART. It sends data to the UART if data is available in any of the buffers. The IRQ enable register of the interrupt controller is activated. The IER and the FIFO control register of the UART are programmed to trigger an interrupt after receiving 4 bytes of serial data. When an interrupt is initiated by the UART, the ARM7TDMI starts its interrupt service routine.

Example 6-7. main_ARM.c

```
#include "interrupt.h"
#include "Bluetooth.h"
#include "devices.h"

int gIntrFlag, pi_flag, gUartRxFlag, gUartTxFlag;
int main(void)
{
 int uart_data = 0;
 int i;
 int size = 0;
 char reason = 0;
  /* Enable IRQ Enable register of Interrupt Controller
*/

outl(IRQ_ENABLE_SET,IRQ_CMR_BIT|IRQ_CI_BIT|IRQ_UART_BIT
);

/* Set UART for word length of 8 bits with 1 stop bit */
        outc(UART_LINE_CNTRL_REG, 0x03);

/* Set Interrupt enable register for receving data */
        outc(UART_INT_ENABLE_REG, 0x01);

/*Set UART to send interrupt on receiving every 4 bytes
*/
        outc(UART_FIFO_CNTRL_REG, 0x41);

/* Baud rate configuration = 115 KB per sec/
UART_CLK_DIV */

        outc(UART_LINE_CNTRL_REG, 0x83);
```

```
        outs(UART_DLL_REG, UART_CLK_DIV);
        outc(UART_LINE_CNTRL_REG, 0x03);
        outc(UART_INT_ENABLE_REG, 0x01);

/*Set Buffer flags to false */
        outl(BUF1_FLAG, 0);
        outl(BUF2_FLAG, 0);

/*Fill in the Buffers */
       for (i=0; i<BUF_SIZE; i++)
       {
       outl(BUF1 + i*4, i+ 10);
       outl(BUF2 + i*4, i+100);
       }

/* Set Buffer flags to true when they are filled */
       outl(BUF1_FLAG, 1);
       outl(BUF2_FLAG, 1);
       outl(BUF1_SIZE, BUF_SIZE);
       outl(BUF2_SIZE, BUF_SIZE);

/* Wait loop for interrupt from UART or BTC */
        while (1) {

/* If data is available to be sent to UART */
          if (gIntrFlag == 1)
           {
            gIntrFlag = 0;
            reason = inl(INTERRUPT);
            switch (reason){
              case BUF1_FULL: size = inl(BUF1_SIZE) ;
                        for (i=0; i< size; i++)
                            uart_data = inl(BUF1 + i*4);
              case BUF2_FULL: size = inl(BUF2_SIZE) ;
                        for (i=0; i< size; i++)
                            uart_data = inl(BUF2 + i*4);
              case NEW :break;
                           }
```

```
            }

/* If data from UART has been read into the Buffer */
           if (gUartRxFlag == 1)
             {
             gUartRxFlag = 0;

/* routine to process the data from UART */
             }
                   }
}
```

Interrupt Handler

The interrupt handler routine is run on the ARM7TDMI processor. Control shifts to this routine whenever there is an interrupt through the hardware RTL to the ARM7TDMI. In the interrupt service routine, the interrupt status is checked to determine the device that has initiated the interrupt. This is done by checking the contents of the ISR of the interrupt controller.

Example 6-8. INT_Handler.c

```
#include <stdio.h>
#include "interrupt.h"
#include "Bluetooth.h"

extern void TMR_INT_Handler(void);
extern void PI_INT_Handler(void);
extern void UART_dev_driver(void);

__irq void INT_Handler()
{
long int IRQ_status;

/* Set interrupt controller status register */
   IRQ_status = inl(IRQ_STATUS);

   if (IRQ_status & IRQ_UART_MDM_BIT)
      UART_dev_driver(); /* ISR for UART*/
```

```
    if (IRQ_status & IRQ_TMR_BIT)
       TMR_INT_Handler();   /* ISR for Timer */
    if (IRQ_status & IRQ_PI_BIT)
       PI_INT_Handler();   /* ISR for Parallel I/O Port */
}
```

UART Device Driver

Upon receiving an interrupt from the UART, the ARM7TDMI executes the device driver routine. All lower priority interrupt devices are masked, and the data is transferred from the UART receive buffer to the memory. The IERs of the UART and the interrupt controller are restored after the data transfer.

Example 6-9. UART_dev_driver.c

```
#include <stdio.h>
#include "interrupt.h"
#include "devices.h"
#include "Bluetooth.h"

extern int gUartRxFlag, gUartTxFlag;
void UART_dev_driver()
{
long int IRQEnable, status, i, low_pr_bits;
char int_status, int_enable, data;
/* Set Interrupt Enable bits for Interrupt Controller */
       IRQEnable = inl(IRQ_ENABLE);
       low_pr_bits = IRQEnable & MASK_PRIORITY_5;
       outl(IRQ_ENABLE_CLR, low_pr_bits);
       status = inl(IRQ_RAW_STATUS) ^ IRQ_UART_BIT ;

/* Disable interrupt from source */
       outl(INTC_LATCH_REG, status);
       int_enable = inc(UART_INT_ENABLE_REG);
       outl(UART_INT_ENABLE_REG, 0x00);

/* Disable source within Interrupt Controller */
       outl(IRQ_ENABLE_CLR, IRQ_UART_BIT);
```

```
/* Mask lower priorities */
      outl(IRQ_ENABLE_SET, MASK_PRIORITY_5);

/* Serve the highest pending interrupt (ISR bit) */
      int_status = inc(UART_INTR_ST_REG);
      int_status &= 0x0f;
      switch (int_status)
        {
           case 0x01: {

/* No interrupt is pending */
/* Transfer data to Memory Buf1 from UART */
                       for (i=0; i<UART_IO_SIZE; i++)
                       {
                         data = inl(UART_RCV_BFR_REG);
                       outl(BUF1 + i*4,data);
                       }

/* Set flag to TRUE when UART receive buff is read */
                         gUartRxFlag = 1;
                         break;
                       }
           case 0x04: {

/* Received data available */
                       for (i=0; i<UART_IO_SIZE; i++)
                       {
                         data = inl(UART_RCV_BFR_REG);
                         outl(0x94080100 + i,data);
                       }
                         gUartRxFlag = 1;
                         break;
                       }
           case 0x02: {

/* Transmitter holding register empty */
                         for (i=10; i<26; i++)
                       {
```

```
                        outl(UART_RCV_BFR_REG, i);
                      }
                        gUartTxFlag = 1;
                        break;}
            default: break;
        }

/* Restore UART Int_enable register */
        outl(UART_INT_ENABLE_REG, int_enable);

/******End of Body******/

/* Restore Enable bits in the end */
        outl(IRQ_ENABLE,IRQEnable);
        outl(IRQ_ENABLE_CLR, 0);

}
```

6.4.4.11 Running the Simulation

In the hardware RTL code, the ARM7TDMI module is instantiated and simulation is run using the UART hardware testbench. The software code runs on the ARM7TDMI processor and executes the interrupt service routines of the various devices connected to the bus. The breakpoints can be set in the software code and the contents of the processor register and debug memory can be examined. The desired hardware signals can be probed and output waveforms can be observed using the hardware simulator.

6.4.4.12 Data Transfer from the UART to Flash Memory

The following is an example sequence of transferring data from the UART to Flash memory.

- The main routine, interrupt handler, and UART device driver run on the processor (ARM7TDMI). The software debugger can be run by setting breakpoints in the source code for detail analysis, if required.
- The hardware testbench is run on the hardware simulator.

- The UART is configured by the processor to receive data from the peripheral connected to it. In this example, output of the peripheral is simulated through the hardware testbench.
- The UART receives data from the peripheral (hardware testbench) and generates an interrupt to the processor whenever it receives 4 bytes of data.
- The processor goes to the interrupt handler routine and finds the cause of the interrupt generated by the UART and executes the UART device driver routine.
- The processor reads the data through the advanced system bus (ASB) to advanced peripheral bus (APB) bridge. It places the data into Flash memory and clears the interrupt flag that was generated by the UART in order to be ready to receive the next data.

Example 6-10 shows a memory dump after the UART data transfer.

Example 6-10. Flash Memory Dump

```
Memory    +0 +1 +2 +3 +4 +5 +6 +7 +8 +9 +a +b +c +d +e +f
940000C0  11 22 33 44 x  x  x  x  x  x  x  x  x  x  x  x
```

Example 6-11 shows the processor register status after the execution of the statement in the main routine. The waveforms on the bus signals and UART can be observed using the GUI provided in the hardware simulator.

Example 6-11. Processor Register Status

```
outc(UART_LINE_CNTRL_REG, 0x03);

Register          = Hexadecimal value
Exec_Addr  = 000000D4
CPSR       = 60000053
R0         = 21C00000
R1         = 94000084
R2         = 00000000
R3         = 00000000
R4         = xxxxxxxx
R5         = xxxxxxxx
R6         = xxxxxxxx
R7         = xxxxxxxx
```

```
R8          = xxxxxxxx
R9          = xxxxxxxx
R10         = xxxxxxxx
R11         = xxxxxxxx
R12         = 00000000
R13         = 9403EFFC
R14         = 00000003
R15         = xxxxxxxx
```

6.4.4.13 Data Transfer from Flash Memory to the UART

The following is an example sequence of transferring data from Flash memory to the UART.

1. The main routine, interrupt handler, and UART device driver run on the processor (ARM7TDMI).
2. The hardware testbench is run on the hardware simulator.
3. The UART is programmed by the processor to transmit data from the processor to the peripheral connected to it. In this example, the peripheral reading data from the UART is simulated through the hardware testbench.
4. The UART generates an interrupt indicating that it is ready to accept the data to be transmitted to the peripheral connected to it. The interrupt is generated to the processor through the interrupt controller.
5. The processor goes to the interrupt handler routine and finds the cause of the interrupt generated by the UART and executes the UART device driver routine.
6. The processor transfers the data from Flash memory to the UART through the ASB/APB bridge. The UART transfers the data to the peripheral. The processor also clears the interrupt flag generated by the UART in order to be ready to transmit the next data.

The waveforms on the bus signals and the UART can be observed using the GUI provided in the hardware simulator. The processor status can be checked by performing a memory dump using the software debugger.

6.5 Rapid Prototype Systems

Rapid prototype systems (RPS) are hardware design representations of the design. The key to successful rapid prototyping is to realize the prototype quickly. Some approaches include emulation and reconfigurable and application specific prototyping systems. In emulation, the target design is mapped into a reconfigurable platform built from array processors or FPGA devices. The prototyping systems are discussed in more detail in this section.

Figure 6-12 shows a simple block diagram of a rapid prototype environment. The processor is represented by a bonded-out chip that interfaces with the source-level debugger through an ICE and the peripherals through the processor bus. The peripherals are represented with either real chips or implemented in FPGAs.

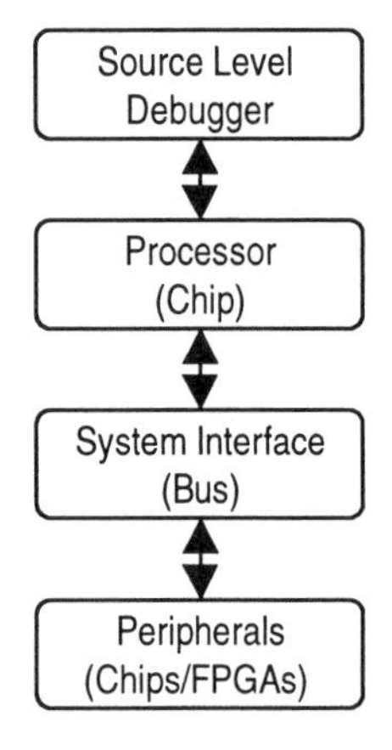

Figure 6-12. Rapid Prototype Environment

Some of the features of RPS are:

- **Applications**: Can configure a wide range of applications for a selected domain. For example, a platform based on an ARM processor can be used for control operations in a cell phone or a Bluetooth device in the wireless application domain.
- **Performance**: The performance obtainable is significantly higher than is achievable with software simulators. It may be possible to run an RPS at real-time speed for some applications.

- **AMS devices**: Supports the integration of analog/mixed signal (AMS) modules, such as analog-to-digital, digital-to-analog converters and radio frequency (RF) modules.
- **ECO**: Allows faster engineering change orders (ECO) for minor design modifications if the systems are based on programmable devices, such as FPGAs.
- **Software**: Software developed for the prototype can be used for the final product integration with few or no modifications.

6.5.1 Limitations of RPS

Some of the limitations of rapid prototypes are:

- **Design partitioning**: Partitioning the design into multiple FPGAs can be a major challenge because of the limitation on the maximum number of pins available in FPGAs (typically 400 to 450 pins). The process of partitioning can take a significant amount of time and should be incorporated in the project schedule.
- **Plug-in modules**: In the case where plug-in modules are not available, they must be developed, which takes a significant time and should be incorporated in the overall project schedule.
- **Design modification**: The design's RTL code might need to be modified to fit the design into the FPGAs efficiently.
- **Interconnect delays**: Signal delays due to interconnection devices reduce overall system performance.

6.5.2 Reconfigurable RPS

In this approach, the target design is mapped to off-the-shelf devices, such as control processors, digital signal processors (DSP), bonded-out cores, and FPGAs. These components are mounted on daughter boards, which plug into a system interconnect motherboard containing custom programmable interconnect devices that model the target system interconnect. In some systems, the interconnects are fixed.

Figure 6-13 shows a simple block diagram of a typical reconfigurable RPS. It consists of a system motherboard that holds the bus interconnection (BIC) devices. It can incorporate the bonded-out cores, such as CPU, memory, and AMS modules. Also, there is a provision to plug-in FPGA modules to implement the IPs for which there are no bonded-out cores available. The system provides connectors for an ICE and LA for debugging. The analog signal measurements can be done using an oscilloscope.

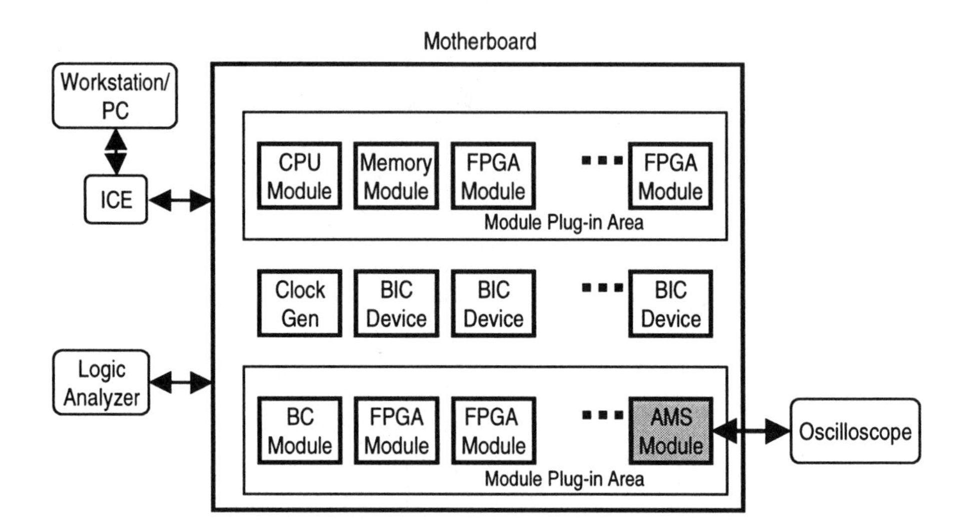

Figure 6-13. Block Diagram of a Reconfigurable RPS

6.5.2.1 Selecting a Reconfigurable RPS

The following areas should be considered when selecting a reconfigurable RPS solution.

- **Availability of plug-in modules**: Should support the processors to be used in the SOC design. For example, ARM, MIPS, and FPGA modules are available from the vendors.
- **AMS devices**: Should provide the facility to plug in AMS devices to realize a complete hardware that represents the SOC. This significantly helps in validating system and improving time-to-market requirements.
- **Configuration facility**: Whether the configuration software offers FPGA implementation. This eliminates design partitioning, synthesis, and preparing a downloadable file. One such tool is Certify.
- **Interconnection facility**: Should be possible to make interconnections among the plug-in modules as per the required bus standards, such as PCI, AMBA, or a user-defined bus.

- **Performance**: Some vendors claim the RPS to be working at 20MHz clock speed.
- **Debug tools interface**: Should provide an easy way of connecting the ICE and LA for debugging.
- **Cost**: Varies depending on the system type, configuration, and so on.

6.5.2.2 Methodology of Reconfigurable RPS

The reconfigurable RPS methodology assumes that the system design is already performed, HW/SW partitioning is complete, and the required set of IP blocks for the SOC are identified. The methodology steps involved in using reconfigurable RPS for SOC verification are as follows.

1. **Select modules**: Select the bonded-out core plug-in modules of the standard devices available from the vendors.
2. **Select FPGAs**: Identify and select the FPGAs to be used in the prototype in case the bonded-out core modules or devices are not available.
3. **Implement FPGAs**: Synthesize, simulate, and place and route the RTL design of the blocks to be implemented in FPGAs.
4. **Fabricate modules**: Fabricate the FPGA and processor plug-in modules using printed-circuit board (PCB) tools by performing schematic entry, place and route. Fabrication of the PCBs and assembly of components can be done by the PCB vendors. The modules are plugged into the system.
5. **Debug**: Connect the processor ICE through the workstation or personal computer. Also connect the LA to the devices or connectors as per the requirement.
6. **Interconnect system bus**: Perform through the programmable switches or fixed connections on the backplane.
7. **Install software**: Install the debugger software into the workstation or PC. Download the software and debug the firmware/application code.
8. **Compare results**: Capture and record the test patterns and compare the recorded results with the expected results.
9. **Make reports**: Make a report of the errors found in the process of verification. Also document the test cases that detected the errors. This helps the development engineers to understand under what circumstances the errors occurred and identify the hardware and software portion of the design that generated the errors.
10. **Fix errors**: Discuss the errors found with the development team and fix the errors in hardware and software accordingly.

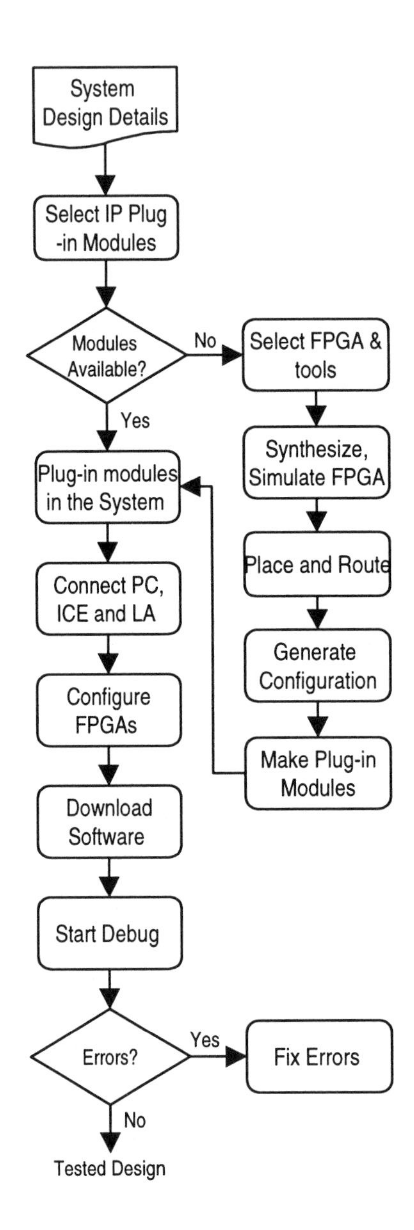

Figure 6-14. Reconfigurable RPS Methodology Flow

6.5.3 Application-specific RPS

The application-specific prototype maps the target design to commercially available components. It has limited expansion and reuse capability. Typically, these prototypes are built around board support packages (BSP) for the embedded processors, with additional components (memories, FPGAs, and cores) added as needed. Depending on the IPs selected for the design, the BSPs can be used to develop and debug the hardware and software before the availability of the SOC. For example, IP providers, such as ARM, DSP Group, MIPS, and Motorola, provide BSPs based around their processor IPs.

BSPs offer the following features:

- Help to understand IP functions, including processors and peripherals, quickly
- Demonstrate the features of the product to be designed
- Minimum configuration is built-in for using the IP
- Available with software development kits consisting of a compiler, assembler, linker, loader, and debugger
- Ability to develop and debug the hardware and software for the intended SOC design based on the processor core used in the BSP
- Can plug in additional IP modules required for the intended application through FPGA implementation
- Facility to probe and monitor pins and signals in the system
- Ability to connect an ICE and LA for debugging

6.5.3.1 Commercially-available BSPs

Some of the BSPs available from processor core vendors are described here. Other processor core and third-party vendors also provide similar BSP solutions.

Microprocessor-based: ARM has a BSP based on advanced microcontroller bus architecture (AMBA). It consists of an ARM7TDMI processor chip, arbiter, address decoder, memory, processor to peripheral bridge, two timers, interrupt controller, two UARTs, parallel port, and two PC card add-in slots. The memory consists of flash/EPROM, SRAM, and DRAM. The BSP has software and an ICE for development and debugging. It operates at 20MHz clock and has a facility to connect the LA.

The BSP module can add external modules by bringing out the ASB and APB signal lines on suitable connectors. The external modules can be designed taking the signal pin details into consideration, enabling direct plug-in to the connectors on the BSP. Refer to www.arm.com for more details.

DSP-based: The DSP Group BSP is based on the OakDSP. It consists of a DSP, address decoder, external memory, an audio Codec, and glue logic. It operates at 40MHz clock speed and has the facility to connect an LA. The debugging feature is provided through a PC add-in module. The BSP has the necessary software and example application programs.

The BSP module is incorporated with a general purpose area for adding any user-specific logic, depending on the application. This capability is provided by bringing out the processor address, data, and control bus signal lines on suitable connectors. Refer to www.dspg.com for more details.

6.5.3.2 Application-specific RPS for the Bluetooth SOC

Figure 6-15 shows a simple block diagram of an application-specific RPS for the Bluetooth SOC that is based on ARM's AMBA based BSP. The BSP consists of all the blocks, except the universal serial bus (USB), Codec, and Bluetooth link controller blocks. The missing blocks can be incorporated in an ASB module and plugged into the connectors provided on the BSP. The Bluetooth link controller and digital portion of the Codec can be implemented in an FPGA. The AMS blocks (analog-to-digital converter (ADC), digital-to-analog converter (DAC), and the USB analog portion devices) can be connected outside the FPGA using standard chips. Refer to www.arm.com for guidelines on developing and incorporating the additional modules to the BSP.

The software tools supplied with the BSP contain a source-level debugger, compiler, assembler, and linker, which can be used for software development and debugging. The application software and firmware can be tested using this application-specific RPS.

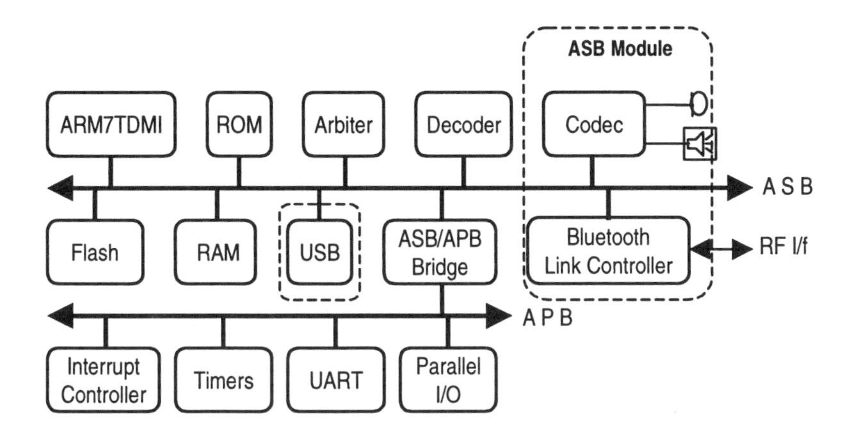

Figure 6-15. Block Diagram of Application-specific RPS

6.5.3.3 Application-specific RPS Methodology

The application-specific RPS methodology assumes that the system design is already performed, HW/SW partitioning is complete, and the required set of IPs for the SOC and the BSP are identified and selected.

Figure 6-16 shows the application-specific RPS methodology flow. The steps involved in this approach for SOC verification are as follows:

1. **Select BSP**: Select and study the BSP for available IPs and map the SOC design.
2. **Select FPGAs**: In case the required IPs are not available, select the FPGAs for implementing the IPs.
3. **Implement FPGAs**: Synthesize, simulate, place and route, and generate the configuration files for the FPGAs, using RTL code as the input. Develop an additional PCB, if required.
4. **Configure BSP**: Configure the BSP as per the design requirements.
5. **Connect modules**: Connect additional modules to the BSP connectors.
6. **Debug**: Connect the ICE through the workstation or PC, and the LA to the devices.
7. **Install software**: Install the software tools on the workstation or PC. Download the software and debug the device driver and application code.

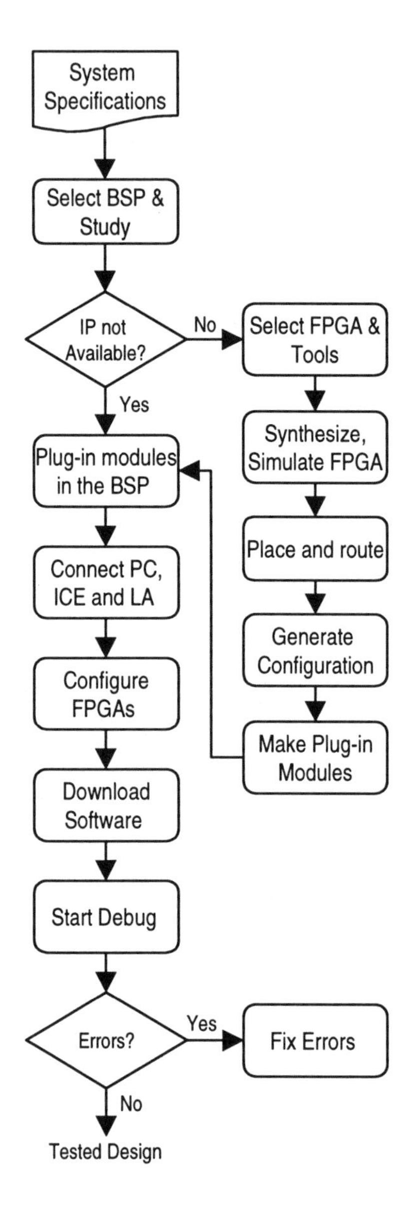

Figure 6-16. Application-specific RPS Methodology Flow

8. **Compile**: Compile and run the software tests.
9. **Compare results**: Capture and record the test patterns and compare the recorded results with the expected results.
10. **Make reports**: Make a report of the errors found. Also document the test cases that detected the errors. This helps the development engineers to understand under what circumstances the errors occurred and identify the hardware and software portion of the design that generated the errors.
11. **Fix errors**: Fix the errors in hardware and software accordingly.

6.5.3.4 Limitations of Application-specific RPS

Some of the limitations of application-specific RPSs are:

- Limited to selected application domains. In some cases, once the product development is completed, the RPS cannot be reused unless the future product is based on the same platform.
- If additional modules are required, the time spent in developing a module can be significant, because it involves FPGA implementation, PCB fabrication, assembly, and module testing.

6.6 Comparing HW/SW Verification Methods

The HW/SW co-verification environments explained in the previous sections are compared in Table 6-1. The areas compared are overall speed, debugging capabilities, software testing level, the timing obtainable, and the cost of the environment. When selecting a verification method, the verification goals, the coverage, accuracy, and performance required, project schedules, and the resources available also need to be taken into account.

Table 6-1. Comparing HW/SW Co-verification Environments

Environment	Speed	Debugging	Software	Timing	Cost
Soft/Virtual prototype	Medium	Algorithm	Firmware	No	Low
Co-verification	Slow	High	Firmware	Yes	High
Rapid prototype system	Medium	Low	Real	Yes	Medium
Emulation	High	Low	Real	Yes	Very high

6.7 FPGA-based Design

Many design houses have been using FPGAs to develop product prototypes quickly to meet time-to-market pressures. In fact, some consumer electronics are available with FPGA implementations. Recent developments in FPGA technology have made high-capacity, high-speed FPGAs available at a lower cost, thus making them suitable for implementing SOC prototypes.

FPGA implementation offers the following advantages over application-specific integrated chips (ASIC).

- **Incremental design**: Capability to develop the IP or system incrementally. The sub-blocks of the design can be tested as and when completed, providing greater confidence in the design and faster ECOs.
- **No risk**: Because FPGAs are reprogrammable, the risk and cost involved when the design is being implemented in ASIC are eliminated.
- **No manufacturing delays**: Does not involve manufacturing, since the design is implemented in a device that is already manufactured as a standard part.
- **No device testing**: No device test, no device sign-off, no non-recurring engineering costs, and no delays involved in developing the FPGA-based implementation.
- **System design**: Early availability of the FPGA-based prototype of a system can greatly assist in system hardware and software design and development.

6.7.1 Guidelines for FPGA-based Designs

Some of the guidelines that could be useful when implementing FPGA-based designs are as follows.

- The basic ASIC design should be structured in hierarchical ASIC blocks will help in fitting in individual FPGAs. This reduces the risk to partition ASIC design blocks among multiple FPGAs.
- Begin the design with conservative capacity and I/O pin usage for a given FPGA so that design changes do not lead to a need to repartition.
- Bring out critical internal signals for monitoring the internal logic operation on the unused pins of FPGA. This helps in identifying the source of errors quickly.
- Keep the number of clocks used in the design to less than or equal to the number of global clocks available in the FPGAs that are selected for the target system/ board implementation.

- For designs that need low capacity buffers, such as first-in-first-out (FIFO), single port memory, and dual port memory, the FPGA internal memory capability can be used. This improves the design's overall speed and reliability.

6.7.2 FPGA-based Design Methodology

The steps involved in FPGA-based design methodology are as follows, as shown in Figure 6-17.

1. **Design**: Input is system specifications. Sometimes the IP blocks are already available. The design can be in HDL, such as Verilog or VHDL. Some FPGA implementation tools allow design entry through schematic capture.
2. **Simulation**: Functional simulation uses the testbenches created. The same testbenches can be used for functional simulation after synthesis. Standard event-based and/or cycle-based simulation tools can be used. If the implementation is complex, then it is recommended to use the same verification methodology as in ASIC or SOC is adopted.
3. **Synthesis**: Involves using standard synthesis tools to translate the RTL code into a gate-level netlist that can be mapped to the FPGA logic blocks. This involves setting the parameters in the synthesis tool as per the deign requirements.
4. **Place and Route**: Chip layout is prepared using the tool supplied by the FPGA vendor.
5. **Timing verification**: Checks for timing violations in the FPGA implementation. Timing checks, include setup and hold time and speed requirements, and is done for all signal paths in the FPGA.
6. **Configuration**: Configuration file used to program the FPGA to implement the intended functionality is generated. The configuration is downloaded into the memory of the FPGA, either through a configuration download cable or a serial PROM connected to the FPGA in the target system or board.
7. **Prototype**: Can be an RPS or a target system/board in which the FPGA is located.
8. **Testing**: Done in-system as per design requirements. Stimulus is applied through the pattern generator, and the signals are captured and recorded using the LA. The recorded results are compared with the expected results for correctness.
9. **ECO**: Implement any design modifications and then repeat all the steps from design entry through the prototype testing.

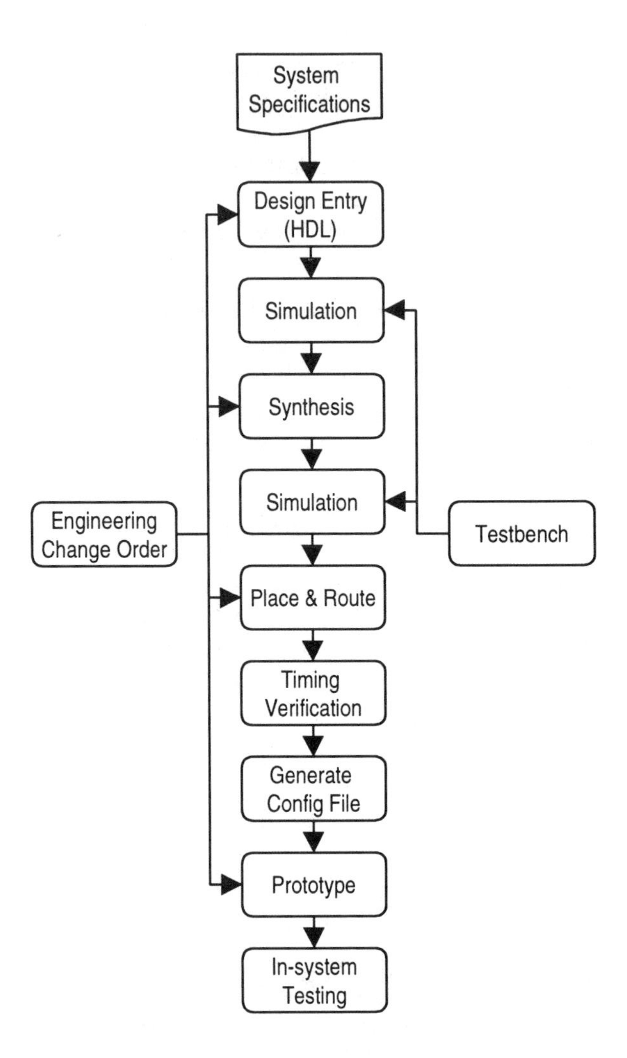

Figure 6-17. FPGA-based Design Methodology Flow

6.8 Developing Printed Circuit Boards

If additional IP blocks are required to interface with the BSP in an application-specific RPS, they are implemented with FPGAs and standard chips. A PCB is used to hold the FPGAs and standard chips. Figure 6-18 shows the PCB development methodology. The following steps are involved in PCB development, as shown in Figure 6-18.

1. **Specify modules**: Module specifications include the logic to be implemented in the PCB, connector-signal electrical and pin details, and mechanical dimensions.
2. **Specify standard chips**: Select standard chips for the IP blocks. If the IP blocks are not available in the library, then implementation of the IP blocks is done in FPGA, using FPGA-based methodology. FPGA pin-out is set and passed to the design entry step.
3. **Design entry**: Uses the schematic capture tools that are embedded in the standard PCB design package. The schematic symbols for the devices are created in case they are not available in the library. Some PCB design packages provide the capability of simulating the design as per the specifications.
4. **Generate netlist**: Uses the packager tools embedded in the PCB design package. The netlist is checked against the schematic that was entered, which helps in checking for any floating nets.
5. **Place and Route**: The mechanical symbols for the components are created if they are not already available in the library. The components are placed and routing is done by defining the number of layers, via pad and hole dimensions, track details, and so on. After routing, the design is cleaned with a glossing process provided. The artwork files and drill detail plot are generated and sent to PCB fabrication.
6. **Fabricate PCB**: This is usually done by the subcontractor.
7. **Assemble PCB**: After assembly, the PCB is inspected for any defects, component mounting, errors, solder shorts, opens, etc.
8. **Connect equipment**: The debugging tools, such as an ICE and LA, the pattern generator, and the power supply are connected.
9. **Test**: The FPGAs are configured by downloading the configuration software through the workstation or PC and the download cable. The software is downloaded and debugging is done. After successful testing, the PCB can be plugged into the BSP.

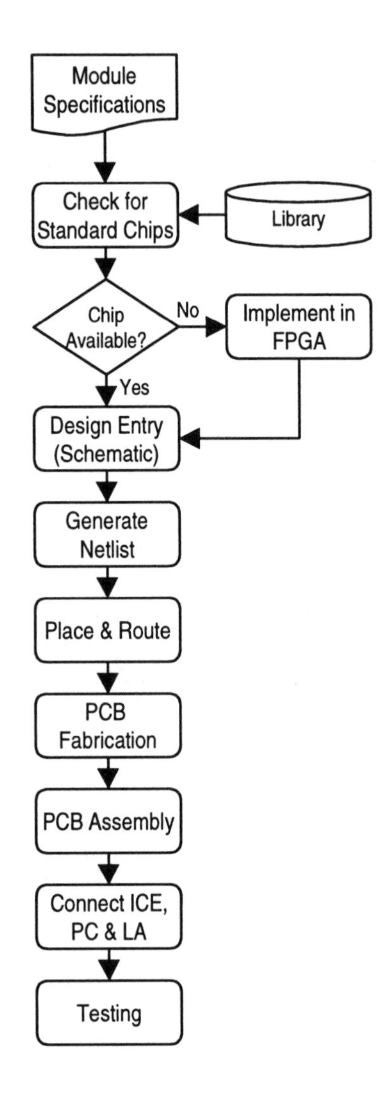

Figure 6-18. PCB Development Methodology

6.9 Software Testing

Software testing is a very critical element to ensure product quality. The increasing amount of software in an embedded system or SOC mandates thorough software testing, which consumes a significant effort in overall product design cycle. The data collected during testing can provide metrics that help in determining system quality.

6.9.1 Software Development Lifecycle

This section briefly describes various software development lifecycle paradigms, with a focus on software testing. Many design houses use the best features of the different software lifecycle models.

6.9.1.1 Waterfall Lifecycle Model

The waterfall lifecycle model, as shown in Figure 6-19, requires a systematic and sequential approach to software development. Following are the details of each phase.

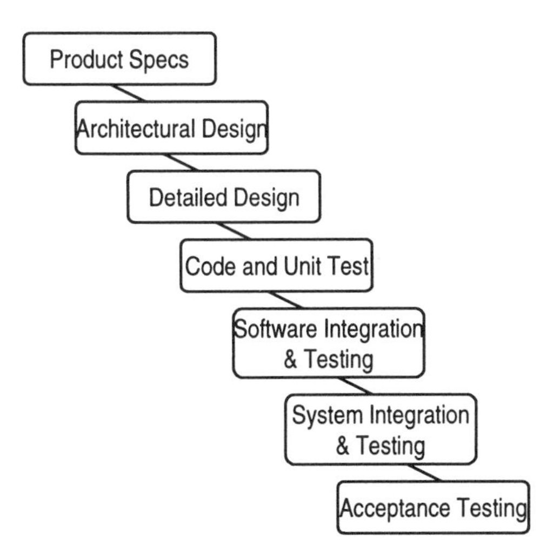

Figure 6-19. Waterfall Lifecycle Model

- **Product specifications**: Specifications are gathered and analyzed, and acceptance criteria for the product are defined.
- **Architectural design**: The architecture and the software components and modules required are defined.
- **Detailed design**: Implementation details of each component and module are specified.
- **Code and unit test**: Detailed coding of each component and module is completed and unit testing is performed to check for the intended functionality. White-box testing methods are used for each module.
- **Software integration**: Software components created by development team members are integrated and tested. There are two main approaches for integration: top-down and bottom-up. In top-down approach, the modules are integrated by moving downward through the hierarchy, starting with the main module.Tests are performed as each module is integrated. The bottom-up approach starts by testing each component and moving upwards.
- **System integration**: The software is integrated with the overall product/system and testing is performed. System testing includes run-time performance testing, stress testing for abnormal situations, security testing, and recovery testing for checking the system in the presence of faults.
- **Acceptance testing**: The tests that are defined in the product specification phase are applied to the product and validated for the intended functionality.

6.9.1.2 V Lifecycle Model

The V lifecycle model, as shown in Figure 6-20, consists of three phases:

- **Architectural**: Consists of product specifications, architectural design, system integration, and acceptance testing.
- **Design**: Detail design and software integration and testing.
- **Implementation**: Includes code design and unit testing of each software component in isolation.

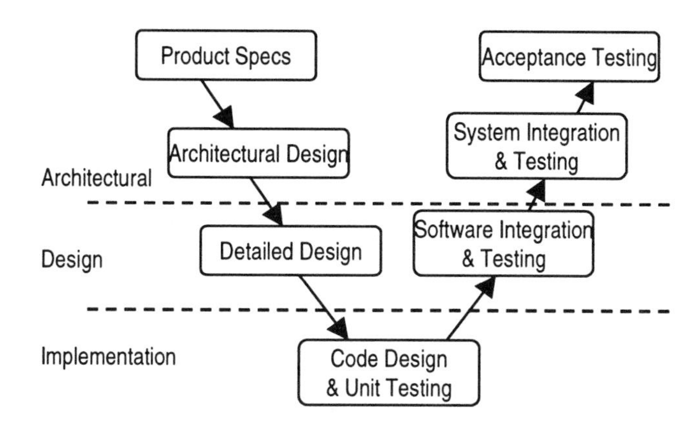

Figure 6-20. V Lifecycle Model

6.9.1.3 Prototyping or Iterative Lifecycle Model

The prototyping (or iterative lifecycle) model approach is best suited in situations where the overall objectives of the software have been identified, but not not identified the detailed input, processing, or output requirements. Other times that it is appropriate is when the development team is not sure of the algorithm efficiency, operating system adaptability, or interface issues.

Figure 6-21 shows the prototyping model. The prototyping starts with gathering user requirements. The development team and the customer discuss and define the high-level objectives for the software. The requirements that are known in the beginning and the areas that require further specification definitions are identified. A quick prototype that represents the already-defined requirements is built and used to further refine the requirements. This leads to iterations as the prototype is progressively tuned to meet the customer's requirements.

The main drawback of the prototyping model is that the customer might see the prototype itself as the product, without considering the software quality. The prototype is developed keeping only the requirements in mind, and not with a focus on software quality. The solution is to define and discuss the objective of developing the prototype in the beginning.

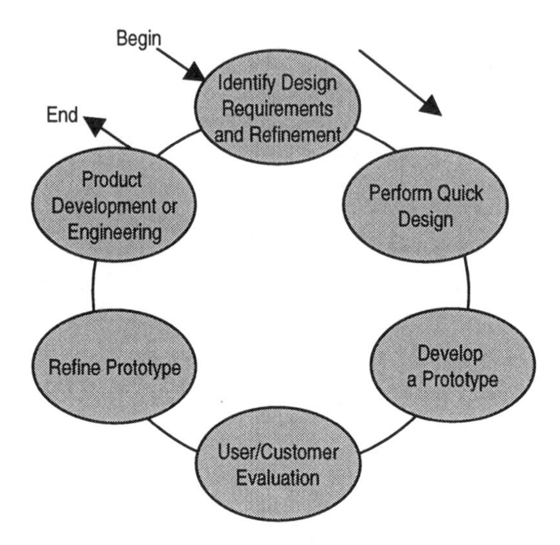

Figure 6-21. Prototype Model

6.9.1.4 Software Maintenance

Software maintenance reapplies each of the steps explained in the lifecycle model to existing software. Maintenance could be required because of software enhancements, modifications, or a new operating system requirements.

6.9.2 Guidelines for Software Development

Following are some guidelines to use for better testability and debugging when developing software for SOC.

- **Program organization**: Maintain a hierarchical program organization that exhibits good use of software program structure.
- **Modularity**: Create modular programs. Partition the software modules into components that perform specific functions. Maintaining modularity helps in identifying errors quickly.
- **Simplicity**: The software should be as simple as possible, since this leads to easier readability, better understanding, and easier testing.

- **Interface**: Create modules with a minimum interface between other modules and the external environment.
- **Self-test features**: Adding a self-test feature to the software helps discover errors during testing.
- **Initialize variables**: Initialize all the variables used in the software. Uninitialized variables can cause non-reproducible or intermittent errors.
- **Comments**: Provide comments wherever required, along with good documentation. This makes code reviews and white-box testing much more efficient.
- **Communication**: Frequent communication is required between hardware, software, and verification teams to better understand system requirements, HW/SW interface requirements, project status, and problems encountered in the development and testing process. The development team members should be available to the verification team so that errors can be understood and fixed quickly.

6.9.3 Software Testing Best Practices

Many design houses use the following best practices for software testing.

- **Specifications**: Early availability of the system specifications helps the verification team to identify the test strategies and environment required to validate the system. Tests are created using the black-box testing method. Creating tests and developing software can take place in parallel, enabling faster integration later. In the design cycle, test metrics can be determined with reference to the specifications.
- **Process interface**: Define the entry and exit criteria for the software process.
- **Automation**: Automating software testing helps minimize manual work and gain higher coverage with greater number of test cases. This involves leveraging already developed tools and scripts, and developing new tools or scripts for tasks that are not automated.
- **Single product and multiple application**: In consumer electronics products, the same basic platform or product is may be used for numerous applications, complicating the testing process. In such a situation, identify the user needs for a particular system configuration, develop the test cases against the identified requirements, and perform testing. This can save a significant amount of time and efforts.
- **Code coverage**: Code coverage measures the elements of the code that have been exercised during testing. Code coverage includes statement, branches, con-

dition, path, and data coverage. Tools with code coverage capabilities are available.

- **Reviews**: Code walk-through and reviews help understand the code and detect errors at the top level. Reviewers should be peers who are not directly involved in developing the software that is under review.
- **Communication**: Frequent communication is required between hardware, software, and verification teams to better understand system requirements, HW/SW interface requirements, project status, and problems encountered in the development and testing process. The development team members should be available to the verification team so that errors can be understood and fixed quickly.

6.9.4 Debugging Tools

Debugging tools are used to identify and isolate errors when integrating hardware and software and to test the firmware and application software of an SOC. Debugging tools offer the following functions:

- Provide access to the processor internal registers
- Control processor execution
- Set breakpoints and single-stepping in the software
- Provide real-time trace
- Do full symbolic debugging
- Ability to perform in-circuit emulation

Debugging tools for SOC, hardware, and software are described in the following sections.

6.9.4.1 Software-based Tools

Software-based debuging tools include ISSs and cycle accurate simulators (CAS). The simulator is a software tool that simulates the functionality of the processor. It interfaces the C models of peripherals and creates a system that represents the target system prior to the availability of real chip or prototype. This allows designers to start developing firmware, device drivers, and application software early. In any application, the simulation speed can never match the actual processor speed. However, these simulators can be used for early development and testing of software, prior to the availability of actual silicon or a prototype.

For an ISS, the processor instructions are simulated with no timing considerations. The ISS executes the software as a sequential list of instructions. It is much faster than a CAS.

With a CAS, the behavior of the processor is modeled to cycle-accurate level. The processor's internal details, such as pipeline, memory interface, and bus protocol, are simulated in greater detail. The timing of the model is precisely the same as the actual processor. The simulation speed that can be obtained is about 1,000 instruction per second. CASs are used in co-verification and chip-level verification.

6.9.4.2 Hardware-based Tools

Hardware debugging tools are useful for debugging the hardware target systems. They include ROM emulators (RE), ICEs, and LAs. Figure 6-22 shows ICE/RE and LA connectivity with a hardware module/system. ICE can be connected to the CPU, and RE to the system ROM.

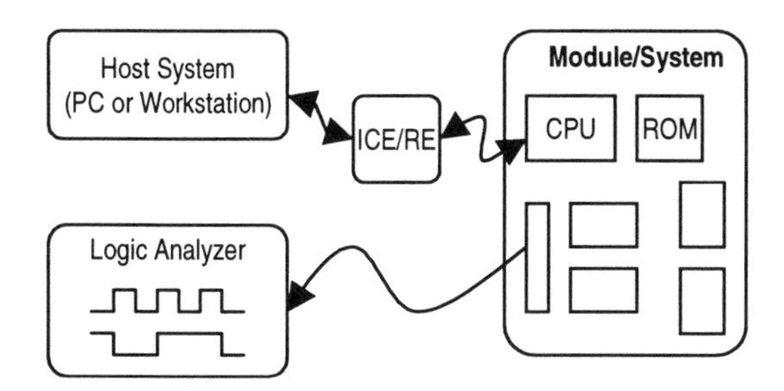

Figure 6-22. ICE/RE and LA Connectivity

An RE plugs into a ROM/Flash socket on the target hardware prototype or system and maps the target ROM to its internal RAM. This facilitates rapid code modification and expedites the debugging process by eliminating the erasable-programmable ROM (EPROM) erase and burn cycles.

ICE is a hardware device that physically replaces and emulates the processor in a target system under test. It plugs into the actual processor socket via a connector pod and emulates the processor exactly to the bus and timing specification level. ICE provides the following capabilities for debugging:

- Accessibility to the memory and registers by allowing read/writes

- Setting the break-points
- Execution control for the processor
- Traces instructions in real-time and the memory and register details with time-stamps
- Ability to profile and take performance measurements of the software routines

The LA is used for high-speed data acquisition in a target system. It provides multiple input channels, signal capture, and intelligent trigger features. The information is displayed relative to a trigger or predefined signal combination, allowing a backward and forward view to help identify the root cause of a problem or event. The signals must be brought out on separate connectors in the target system to connect the LA. Some LAs support the symbolic debugger and disassembler of standard processors, enabling a faster debugging process.

6.9.4.3 Debugging Techniques for SOC

The processor cores embedded in an SOC are not visible, because there are no pins available to connect an ICE and LA for debugging and analysis. The ICE and LA require address, data, and control bus for debugging, but these signals are hidden in an SOC. This requires new techniques and debugging tools to address the needs of the embedded cores. In the last few years, two techniques have emerged for debugging embedded cores: background debug mode (BDM) and a scan-based emulation technique based on the Joint Test Access Group (JTAG) IEEE 1149.1 standard.

BDM is incorporated in Motorola microcontrollers. This is achieved by adding a small debugging logic and additional microcode in the microcontroller. This mode enables an external host processor to control a microcontroller-based target system and access its internal registers and memory through a serial interface. The connectivity between the host processor and the target system is done with a BDM cable. The cable contains a header that connects to the target system, and a parallel port connector for the host. The host can read/write registers and memory, control execution, and reset and restart the target system processor without affecting normal operation. The BDM technique does not require any target processor resources, such as on-chip memory, timers, or I/O pins.

The JTAG IEEE 1149.1 standard was initially used for chip manufacturing testing through boundary scan. The standard also allows performing internal device testing, such as automatic test pattern generation (ATPG) and built-in-self-test (BIST). The standard supports five signals for test implementation: test mode select (TMS), test clock (TCK), test data input (TDI), test data output (TDO), and test reset

(TRST). The core to be tested in a chip is first isolated from other cores, the test input is applied serially through TDI with reference to clock (TCK), and the TDO is sampled and tested.

Many core providers and semiconductor companies have used the JTAG standard for emulation by adding debugging logic to tap the processor's internal registers, instruction, and data bus contents. It is also possible to set breakpoints and control the execution of the software. Processor cores, such as ARM, MIPS, and PowerPC, have JTAG-based emulation features.

The Nexus 5001 Forum was formed in 1998 to define and create a debugging interface standard for embedded processors. The standard is processor- and architecture-neutral and supports multicore and multiprocessor designs. The forum consists of processor vendors, tool developers, and instruments manufacturers. Nexus 5001 is officially called the IEEE-ISTO 5001 standard. For more information on this standard, refer to www.ieee-isto.org/Nexus5001/.

6.9.5 Debugging Interrupts

In embedded system, interrupts cause the processor to suspend executing the current task and execute a different task that is caused by the interrupting device. The interrupt is generated by the device, indicating that the processor's attention is required to handle the data transfer. The interrupt is asynchronous and can occur at any time. Some examples of interrupts are:

- The UART generates an interrupt to the processor whenever it receives a byte of data, requesting the processor to read the byte.
- A timer generates an interrupt to the processor indicating completion of a particular task after the programmed time is elapsed.
- A transmitting device generates an interrupt to the processor indicating it is ready to accept the data from the processor to transmit.

Interrupts can be level- or edge-sensitive. A level-sensitive interrupt is recognized when it is in the active state. It is expected to return to inactive state when it is serviced. An edge-sensitive interrupt is recognized on the transition from the inactive to active state. The interrupt is caused by the transition and not the level.

Figure 6-23 shows the components of interrupt hardware and software.

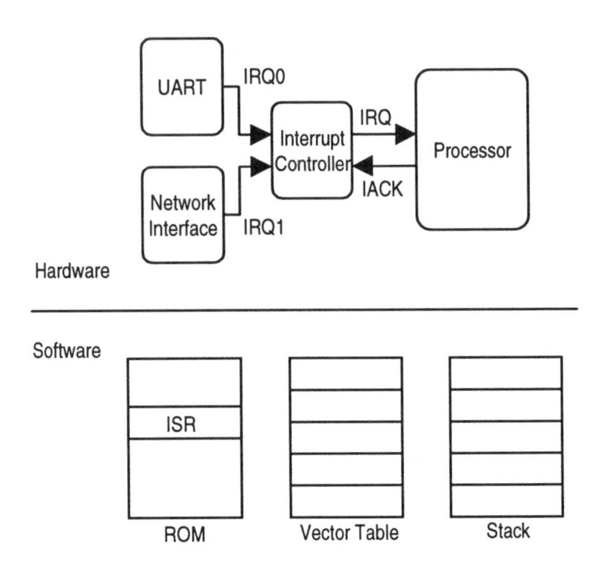

Figure 6-23. Interrupt Hardware and Software

The following guidelines can be used to develop an interrupt service routine (ISR), which helps in debugging the interrupts in an SOC.

- **Interrupt service routine**: Write simple and faster ISRs. Simple ISRs are easier to understand; faster ISRs help in handling all the other interrupts in the system to meet the intended timing requirements. It is recommended to place the critical part of the code in the ISR; any data-processing task should be done in another task. This reduces the overhead on time taken to execute the ISR. For example, following are two situations for making an ISR faster:

  - For a real-time clock interrupt, the ISR should only increment the counter, and a separate routine should handle the time, day, month, and year calculation task.

  - In a target system that processes a command from a host machine serial port, ISR can place the received data into a buffer, and a separate routine can process the command.

- **Stack handling**: Stack is a portion of system memory that is used for storing the return addresses and status of the processor registers when it is interrupted. The system can fail if the stack is not handled properly. It is recommended to allocate a stack size of 2X or 3X times the actual required size. The actual stack size required for the system can be determined by inserting a small chunk of code into an ISR to monitor the stack. This code compares the stack pointer to a limit that is set by the programmer.

- **ISR re-entrancy**: ISR can be a re-entrant routine. This means, while ISR is being executed, it can be recalled by itself or by some other routine. It is required to check that the average interrupt rate is such that the ISR will return more often than it is recalled. Otherwise, the stack overflows and the system fails.
- **Missing interrupts**: Missing interrupts can lead to system failure if not handled properly. To avoid missing interrupts, use a counter or flag in the ISR. The counter is incremented when the interrupt occurs and is decremented when the interrupt is serviced. The counter value of 0 indicates no pending interrupts. If the counter contains a value other than 0, the pending interrupt needs to be serviced.

Summary

Co-verifying hardware and software in an SOC device calls for new methodologies and tools. The amount of software in SOC devices is also increasing, so software debugging early in design development is even more critical.

References

1. Morasse Bob. Co-verification and system abstract level, Embedded Systems Programming, June 2000.

2. Tuck Barbara. The hardware/software coverification challenge, Computer Design, April 1998.

3. Rompaey Karl Van. A checklist for SOC hardware/software codesign, Electronic Systems, April 1999.

4. Foote Andy. Hardware/software co-development: A software engineer's perspective, Integrated System Design, April 2000.

5. Albrecht Thomas W, Notbaur Johann, Rohringer Stefan, HW/SW Coverification performance estimation and benchmark for a 24 Embedded RISC core design, 35th Design Automation Conference, 1998.

6. Seamless CVE, Hardware/software co-verification technology, www.mentor.com.

7. Post Guido, Muller Andrea, Grotker Thorsten. A system-level co-verification environment for ATM hardware design, Proceedings of DATE'98.

8. Dreike Phil, McCoy James. Co-simulating software and hardware in embedded systems, Embedded Systems Programming, June 1997.

9. Berger Arnold S. Co-verification handles more complex embedded systems, Supplement to Electronic Design, March 9, 1998.

10. Leef Serge. Hardware and software co-verification - Key to co-design, Supplement to Electronic Design, March 9, 1998.

11. Zach Gregor, Wilson John. An evolution in system design and verification, Integrated System Design, March 1996.

12. Design environment for a system-on-a-chip, a technical paper, www.synopsys.com.

13. Co-simulation and verification for the embedded software engineer, Application note, www.motorola.com.

14. Kenney Jim, Leef Serge. Speeding a design to market with co-verification, Wireless Systems Design, November 1997.

15. Singletary Alan. Run it first, then build it - Core emulation in IBM microelectronics, IBM-MicroNews Vol. 4, No. 1, First quarter, 1998.

16. Glaser Steve, Evans Ed. Hardware/software co-development and SOC verification, Electronics Engineer, September 1999.

17. Slomka Frank, Dorfel Matthias, ... Hardware/software codesign and rapid prototyping of embedded systems, IEEE Design and test of computers, April-June 2000.

18. Schulz Steven E. Modeling issues for co-verification, Integrated System Design, August 1995.

19. Schulz Steven E. Co-verification strategies in hardware-software co-design, Integrated System Design, August 1995.

20. Mittag Larry. Focus report: Embedded-software development tools, Integrated System Design, May 1997.

21. Shieh Eric. Reconfigurable computing accelerates verification, Integrated System Design, January 2000.

22. Hammer 50/32 accelerator system details, www.tharas.com.

23. RAVE prototyper system data sheet, www.simutech.com.

24. Virtual system prototyping, www.mentor.com.

25. Ryherd Eric. Prototyping embedded microcontrollers in FPGAs, Embedded Systems Conference, Fall 1998.

26. Zak Ralph. A new approach to system level verification, www.aptix.com.

27. Block based prototyping methodology white paper, www.aptix.com.

28. Using emulation for ASIC design in a satellite-based communication system, Design Supercon '97.

29. Browne Jack. Tools take aim at system-level verification, Wireless Systems Design, June 2000.

30. Li Alvin. System verification: essential for digital wireless system-on-chip (SOC) designs, www.aptix.com.

31. Kresta Dave, Johnson Tony. FPGA High-level design methodology comes into its own, Electronic Design, June 14, 1999.

32. Zeidman Bob. An introduction to FPGA design, Embedded Systems Conference 1999.

33. Zilmer Morten, Jensen Peter. MIPS EJTAG on-chip debug solution with complex break and real-time PC trace, www.lsil.com.

34. Designing complex embedded systems without an emulator, www.sdsi.com.

35. Software debugging in high-performance embedded systems, www.sdsi.com.

36. Software debug options on ASIC cores, Embedded Systems Programming '97.

37. Trends in debugging, Embedded Systems Programming '99.

38. Ganssle Jack G. The state of the art of debuggers, Embedded Systems Programming, January 1999.

39. Nath Manju NS. On-chip debugging reaches a nexus, EDN, May 11, 2000.

40. Ryherd Eric. Software debuging on a single-chip system, Embedded Systems Programming, March 1998.

41. Stewart David B. 30 pitfalls for real-time software developers, Embedded Systems Programming, October1999.

42. Howard Scott. A background debugging mode driver package for modular microcontrollers, Application note AN1230/D, www.motorola.com.

43. Jain Prem P, Ali Mohammad Saleem. Pre-silicon embedded software verification using a virtual prototype, Embedded Systems Conference, 1999.

44. Vink Gerard. Trends in debugging technology, Embedded Systems Conference, 1999.

45. Siyami Aiamak. Debugging tools, trends and tradeoffs in an embedded design project , Embedded Systems Conference, 1999.

46. The ARMulator, Application note 32, www.arm.com.

47. Whittaker James A. What is software testing? and why is it so hard?, IEEE Software January/February 2000.

48. Bhagat Robin. Software design methodology for system-on-chip, www.palmchip.com.

49. Peters Kenneth H. Softwre development and debug for system-on-a-chip, Embedded Systems Conference, Spring 1999.

50. Berger Arnie, Payne Jeff. Software development and debug methods for systems-on-silicon, Embedded Systems Conference, Spring 1999.

51. Berge Jean-Michel, Levia Oz, Rouillard Jacques. Hardware/Software Co-Design and Co-Verification : Current Issues in Electronic Modeling, Kluwer Academic Publishers, 1997.

52. Myers Glenford J. Art of Software Testing, John Wiley & Sons, 1979.

53. Kit Edward, Finzi Susannah.Software Testing in the Real World, Addison Wesley, 1995.

54. Hetzel, William. The Complete Guide to Software Testing, Second Edition, John Wiley & Sons, 1993.

CHAPTER 7

Static Netlist Verification

In a system-on-a-chip (SOC) hardware design, the register-transfer level (RTL) code is verified within the simulation environment. The RTL code undergoes several transformations as it goes through synthesis, optimization, scan-chain insertion, clock-tree insertion and synthesis, manual edits, and layout generation. During each transformation, the design's logical behavior needs to be verified for its intended functionality. Historically, the verification has been done with logic simulation, ensuring that the design functions correctly for the given sets of stimulus. However, the size and complexity of SOCs, as well as the exponential increase in the number of vectors to test, results in time-consuming simulations, which affect project schedules. Also, it is extremely difficult to exhaustively verify that the design functions correctly for all possible inputs. This has created a need for efficient verification techniques and methodologies. This chapter addresses the following topics:

- Netlist verification
- Formal equivalence checking
- Static timing verification

Formal equivalence checking and static timing verification are illustrated with reference to the arbiter block used in the Bluetooth SOC design example.

7.1 Netlist Verification

In functional simulation, a design netlist is usually compared to its specifications, using a testbench consisting of a complete set of test vectors. The design is verified against this reference (golden) set of test vectors, as shown in Figure 7-1.

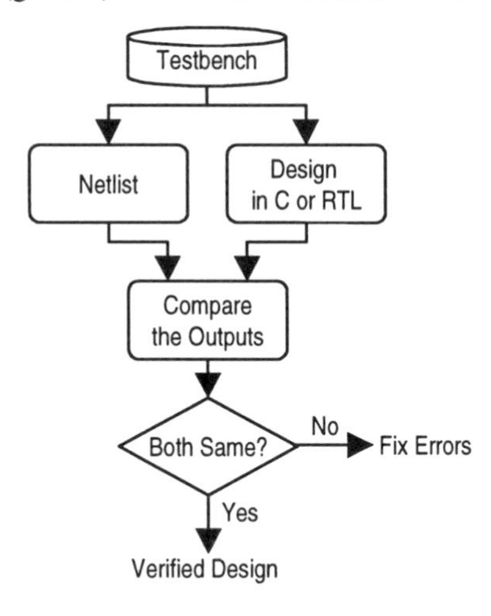

Figure 7-1. Traditional Netlist Functional Simulation

The input design is described in C, Verilog, or VHDL and the final implementation is a netlist. To perform functional simulation, the RTL and the netlist are compiled into simulation models. The golden test vectors are applied as stimuli to both the RTL and netlist, and the resulting outputs are compared. If the outputs match, then the two views of the design are assumed to be functionally equivalent. If mismatches are detected then the source of the mismatch must be indetified and fixed. However, any functionality not addressed by the vectors can cause the simulation to miss errors.

Because of the growing complexity of application specific integrated circuits (ASIC) and SOC, the functional simulation process is cumbersome and time-consuming. Creating testbenches that cover the complete functionality of the design is extremely difficult. Depending on the number of inputs and the internal state of the design, the number of vector sets required increases exponentially. Also, a large number of vector sets is required to cover all the corner cases. Any change in the RTL design leads to updating and resimulating the testbenches.

To overcome the limitations of functional simulation, static verification techniques, such as formal equivalence checking (EC) and static timing verification (STV), have emerged in the last few years. These techniques do not require test vectors to verify a design and take considerably less time to run than functional simulation.

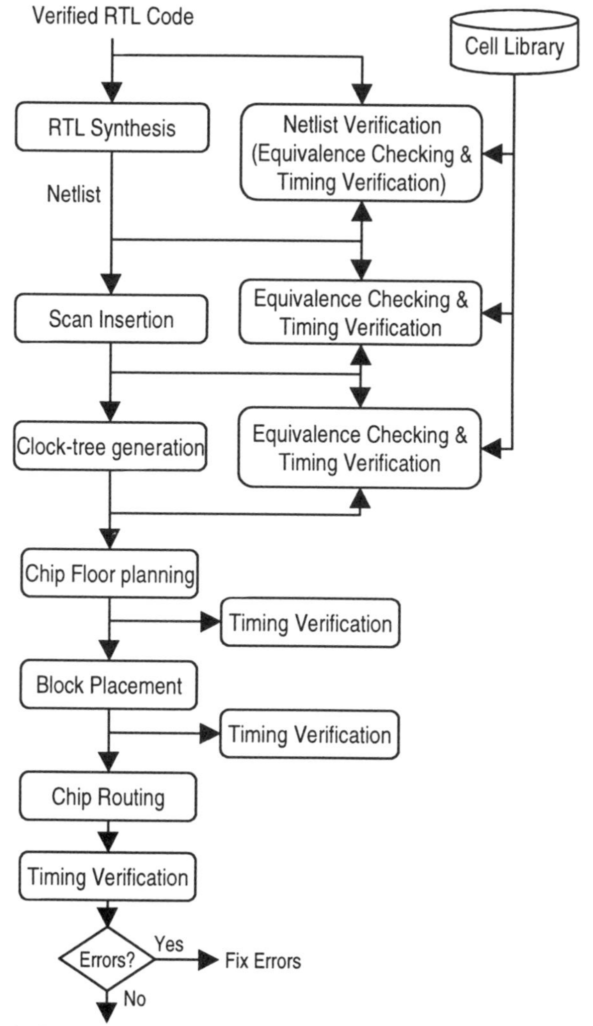

Figure 7-2. SOC Back-end Methodology Flow

Figure 7-2 shows a typical SOC back-end methodology flow. It assumes the already verified RTL design as the golden input. The RTL code is synthesized and the netlist is generated. The netlist verification is performed using EC and STV tools. The necessary cell library is supplied to the tools. The netlist goes through transformation after scan-insertion and clock-tree generation. EC and STV are performed after each step to check for the correctness of the netlist. The floorplan, block placement, and complete chip routing are performed. During each step, STV checks whether the design meets the intended timing requirements. After chip routing, the design is used as input for physical verification and device test steps.

7.2 Bluetooth SOC Arbiter

Figure 7-3 shows a simple block diagram of the example Bluetooth SOC design. The arbiter block is used in this chapter as an example to illustrate the EC and STV methodologies. For a description of the arbiter, see Section 3.3.1 on page 71. Example 3-1 in Chapter 3 shows the RTL code for the arbiter block.

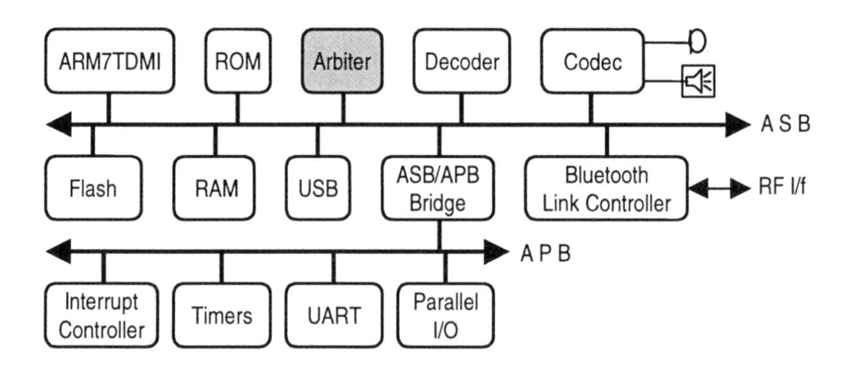

Figure 7-3. Block Diagram of Bluetooth SOC

7.3 Equivalence Checking

Equivalence checkers can easily verify million-gate designs, using less processor time and memory than functional simulation. EC uses a combination of binary decision diagrams and tightly interacting algorithms, called solvers, to prove that two design views are equivalent. EC is successfully used in many design houses for verification.

Typically, the two main steps in EC are:

- Determining the points in the two designs to be compared, called compare points. Compare points can be ports, state bits, or the internal nets of the design.
- Verifying that the logic functions between compare points are functionally equivalent.

EC can be used when a representation of the design has already been verified. It cannot be used for verifying the functionality of the design against the design intent. EC can compare the following views of a design:

- RTL to RTL
- RTL to gate-level netlist
- Gate-level netlist to gate-level netlist

EC assumes that a golden RTL exists and that the RTL has been verified to satisfy the intended system specification through simulation, formal techniques, or both, as shown in Figure 7-4.

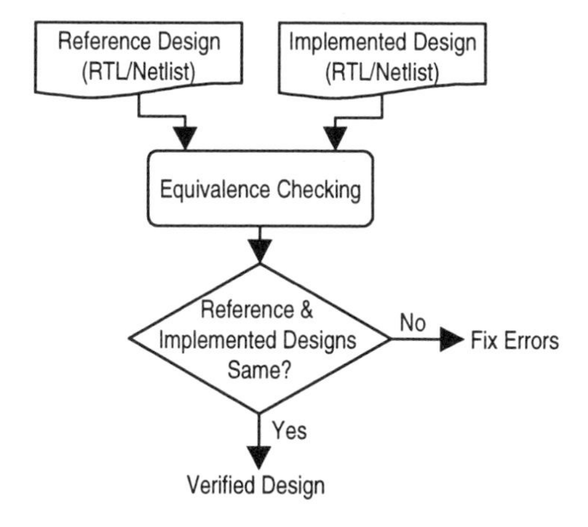

Figure 7-4. Equivalence Checking

As the design progresses through the various transformations, EC only verifies the netlist views generated from the golden RTL. If EC finds that the two views of a design are not logically equivalent, it produces counterexamples to highlight the discrepancies. The source of the differences is then located and corrected.

EC has the following features:

- No vectors or testbench required for verification
- Verifies design from RTL to gate-level, and gate-level to gate-level
- Capacity to handle large designs
- In a hierarchical design, verification can be performed on modified blocks instead of the complete chip, thereby saving a significant time
- Eliminates gate-level simulation, which would take a long time for complex designs
- Allows quick identification of any logic errors
- Significantly shortens design iteration time, allowing multiple design iterations as compared to gate-level functional simulation
- Increases confidence in the design due to the exhaustive nature of the tool
- Improves time-to-market goals by reducing the design cycle time

7.3.1 Selecting an EC Solution

Some of the areas to be considered when selecting an EC solution are:

- **Familiarity**: If formal verification is new to the verification team, they should familiarize themselves with the tools and vendors to determine if this is the appropriate method to use.
- **Capacity**: What design complexity can the tool handle, and the amount of memory, disk space, and compile time required.
- **Set-up time**: How fast the environment can be set up and start performing verification.
- **Performance**: A primary and critical parameter, since most users find performance the major bottleneck in project schedules. Performance of the EC tool depends on design size, compile time, execution time, and host memory.
- **Design representation**: Does the environment adequately represent the design under test, including node states, clocking scheme, combinatorial and sequential logic.

- **User friendliness**: How easy is it to use and how does it fit into the existing chip design methodology.
- **Debugging capability**: Debugging can be usually done using the schematic views and report files.
- **Design environment interface**: Should have easy interfaces with hardware accelerators, emulators, modelers, and libraries to support the overall design methodology.
- **Support**: Check for complete documentation, technical support, online access for downloading and updating programs, and training.

7.3.1.1 Limitations of EC

The main limitations when using EC involve timing and debugging. EC verifies the functionality (logical equivalence) of the design only and does not address timing issues. Timing verification needs to be performed by a separate static timing analysis tool.

Most available EC tools generate reports, which list signals that are different, along with the reason for the difference. It can be cumbersome and time-consuming to find the exact cause of the errors, and verification teams often perform functional simulation to find the cause.

7.3.2 EC Methodology

The following steps are involved in EC, as shown in Figure 7-5.

1. **Prepare design**: Prepare the design as per the guidelines in the requirements.
2. **Compile design**: The compilation of the reference design and implemented design are done using the cell library as the input.
3. **Run EC**: Run the EC tool by configuring it as per the design requirements.
4. **Debug**: Check whether the implemented design is logically equivalent to the reference design. This can be done by going through the report files that are generated by the EC tool. Analyze the results and fix any errors in the design by checking the difference points. Also, the counterexample can be viewed through the schematic view.

Some of the guidelines for performing EC efficiently are as follows.

- **Similar structure**: EC works more efficiently when the designs to be compared are structurally similar. For this reason, run EC frequently between successive transformations where the changes between transformations are minimal.

  In addition, errors introduced at a particular step tend to propagate farther from the source, making the debugging process more difficult. Problems are easier to debug when the two versions are very similar. Usually, the sum total of the time required to verify the individual steps in the design is smaller than the verification between the first and last steps of the design flow when these two steps are far apart.

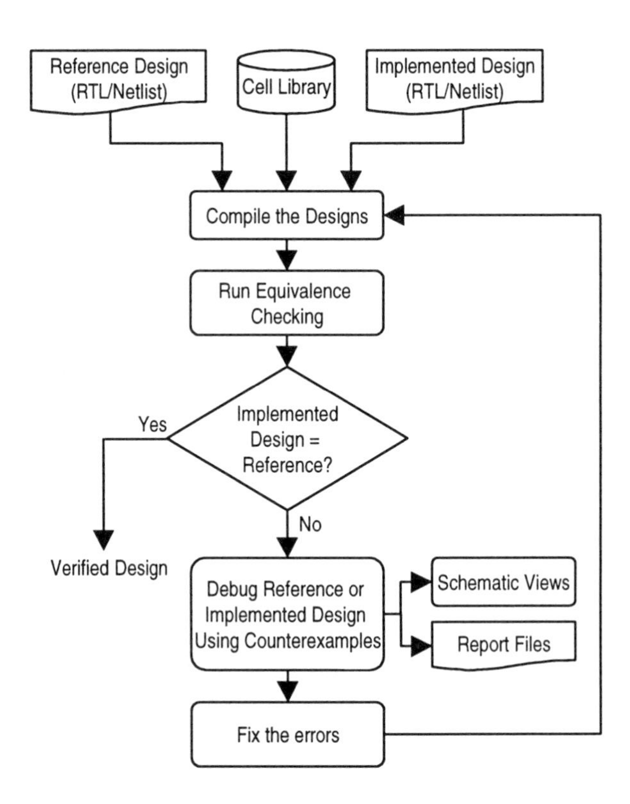

Figure 7-5. Equivalence Checking Methodology

- **Late bug fixing**: Sometimes the RTL needs to be compared to the final gate-level netlist. When bugs are discovered late in the design cycle, the designer might need to recode and resimulate the RTL to get a golden version. In this case, front-to-back EC is required. Even under these circumstances, improved performance and capacity results can justify the extra manual editing of the post-synthesis netlist for gate-gate EC, where the edited post-synthesis netlist is used as a golden reference. Care needs to be taken to avoid introducing new errors during this manual editing process.
- **Bottom-up approach**: Adoption of a bottom-up hierarchical EC methodology helps in tackling performance and capacity constraints for large designs, isolating errors to lower-level blocks, and detecting errors in the early phase of the design. When lower-level blocks are verified, they can be treated as black boxes when the next level of design is verified. At the next level, only the interconnection of the lower-level blocks needs to be verified. This eliminates replicating the cost and time for verifying the lower-level blocks. Most equivalence checkers support hierarchical verification.

7.3.3 RTL to RTL Verification

RTL to RTL verification is generally performed to verify the non-functional changes to the design. This includes partitioning the modules to achieve optimized synthesis and faster speed. It is recommended to perform EC on a module by module basis instead of at chip level.

7.3.4 RTL to Gate-Level Netlist Verification

The equivalence checker converts an RTL representation of the design to an internal representation consisting of sequential devices and combinational logic, essentially performing a synthesis operation on the input. To perform this synthesis, the RTL statements are interpreted according to some well-established guidelines. This interpretation of an RTL statement by the equivalence checker might be different from that by a synthesis tool, because hardware description languages (HDL), such as Verilog, are not always precise in their description of the hardware. An equivalence checker might report that the RTL as not functionally equivalent to the synthesized netlist. An equivalence checker might share the data representation and data interpretation consistent with a specific synthesis tool to eliminate this problem. A bug in the synthesis tool's translation from RTL might not be discovered by the equivalence checker if the same bug existed in its translation.

Some synthesis-specific issues that can cause problems for RTL-gate EC are described below.

Synthesis pragmas are used where specific aspects of the Verilog translation process can be controlled by special comments inside the RTL, such as the `full_case` and `parallel_case` pragmas from Synopsys.

The `parallel_case` directive affects the logic generated by a case statement. In some situations, a case statement in Verilog generates a priority encoder. If a designer chooses not to build a priority encoder, the `parallel_case` directive is used to build multiplexor logic. The following example illustrates using a `parallel_case` directive to encode the states of a state machine using one-hot encoding:

```
reg [3:0] current_state, next_state;
parameter state0 = 4'b0001,
   state1 = 4'b0010,
   state2 = 4'b0100,
   state3 = 4'b1000;
case ( current_state )
   state0 : next_state = state2;
   state1 : next_state = state3;
   state2 : next_state = state0;
   state3 : next_state = state1;
endcase
```

If the above example is implemented as a priority encoder, the generated logic will be very complex. To treat all cases as parallel, the `parallel_case` directive can be used after the case expression, as follows:

```
case (current_state)  //synopsys parallel_case
   state0 : next_state = state2;
   state1 : next_state = state3;
   state2 : next_state = state0;
   state3 : next_state = state1;
endcase
```

The `full_case` directive is used to add "don't care" conditions to non-fully specified case statements. For example, in the following, if the equivalence checker does not understand the `full_case` directive, it does not automatically extract `b = 1'bx when a is 1'b1.`

```
case (a) // synopsys full_case
   1'b0: b =1'b0;
endcase
```

In this case, the condition must be as follows:

```
case (a)
   1'b0: b = 1'b0;
   default:: b = 1'bx;
endcase
```

When a latch input is connected to a constant input of 1 or 0, some synthesis tools propagate the constant across the latch, eliminating the latch. If the equivalence checker does not perform constant propagation across a latch during RTL translation, the result is a mismatch in the number of latches between the RTL and the synthesized netlist. One way to correct this is to turn off constant propagation during synthesis for EC purposes. In this case, the synthesized netlist view being compared to the RTL is not the same as the one implemented in the hardware.

During the synthesis optimization process, logic can be moved across latches to balance the logic between the latches and improve performance. Because equivalence checkers compare the logic between corresponding latches in the RTL and the synthesized netlist, this results in mismatches. Some equivalence checkers handle simple situations of inverters moved across latch boundaries, but most of them have trouble handling the re-timing of complex logic without some manual intervention.

When two state machines have the same encoding, proving their equivalence is straight-forward. However, a synthesis tool can use state encoding to optimize the logic generated from RTL, which is different from the default encoding specified in the RTL. Unless this information is passed to the equivalence checker, the comparison will fail. Because of the exponential complexity posed by proving the equivalence of two finite state machines with different encoding, a designer must input the information about the encoding used during synthesis, or the synthesis tool must generate the appropriate information regarding the encoding used so that this information can be read by the equivalence checker.

It is possible to exclude certain portions of a design that do not require equivalence verification, such as cache and on-chip system memory, which are usually created by a memory generator and might not require proof of equivalence verification. Because the gate-level representations of these elements are usually quite large, many verification cycles might be required, without providing much benefit.

Large multipliers with different architectures can also be excluded from EC. Equivalence checkers have problems proving equivalence of multipliers with different architectures. For these situations, equivalence checkers provide a black-box facility where the design portion to be excluded can be specified as a black box, and mapping between the black-box inputs and outputs of the RTL and the synthesized netlist can be specified. The equivalence checker then verifies the logic at the inputs of the black box and the logic driven by the black-box outputs. The internals of a black box are verified by simulation.

7.3.5 Gate-Level Netlist to Gate-Level Netlist Verification

Once the synthesized netlist has been proven to be functionally equivalent to the RTL representation, successive equivalence checks are run between the various transformations of the synthesized netlist. For the best performance and capacity results, versions of the design that are adjacent in the design flow should be compared. The following items identify typical gate-to-gate EC scenarios.

- **Pre-scan and post-scan insertion**: Inserting scan logic changes the functionality of the design. To perform EC between pre-scan and post-scan gate-level netlists, disable the test logic by applying suitable constant values at the test-enable inputs. This requires complete knowledge of the test logic introduced for proper disabling of the test logic. Simulation can be used to verify that test logic functions correctly, scan chain connections are correct, and test control logic properly drives the scan logic.

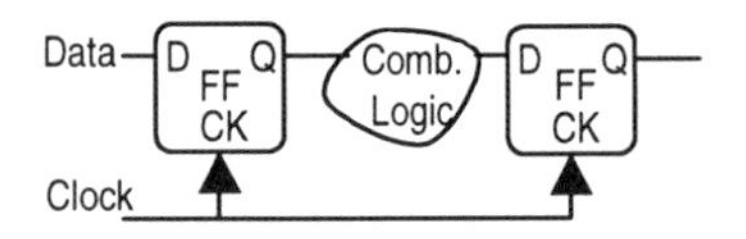

A. Pre-scan Insertion.

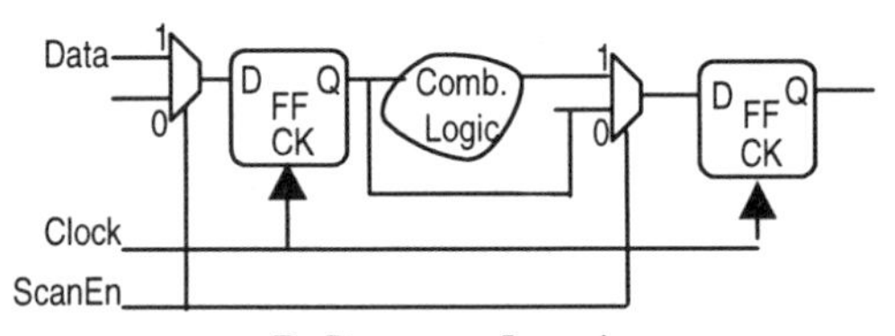

B. Post-scan Insertion.

Figure 7-6. Scan Insertion

Figure 7-6 shows an example of a pre-scan and post-scan circuit. In normal mode, the input ScanEn is set to 1; in test mode, it is set to 0. When the equivalence checker is run on this circuit to compare the pre-scan and post-scan insertion networks, ScanEn input is forced to 1 prior to running EC.

- **Clock-tree synthesis**: Usually clock-tree synthesis results in a flattened netlist, so a designer might need to perform a hierarchical-to-flat netlist comparison. Equivalence checkers usually handle this by flattening the pre-clock-tree synthesis netlist automatically. However, time can be saved by specifying the flattening of the hierarchical netlist before comparison. If the clock-tree synthesis works hierarchically, extra ports may be created at module boundaries because of the need to have both a clock and its inversion available inside a module. This does not pose a problem if the EC is performed at the top-level module. If the verification is to be done at the level of individual modules, mapping the clock ports in the two versions of the design must be set up to ensure that the inverted and non-inverted clock ports are properly correlated.
- **Pre-layout to post-layout netlist**: In this case, the post-layout netlist can be flattened, while the pre-layout netlist is hierarchical. Also, the post-layout netlist can be in a format different than the pre-layout netlist, and might require a format translation if the equivalence checker does not handle that format.

7.3.6 Debugging

When an equivalence checker discovers that outputs or state bits in the two versions being compared are not functionally equivalent, it produces a counterexample illustrating the difference. The counterexample typically consists of the following names and values.

- Comparison points that differ
- Inputs of the logic cone that drive the comparison points
- Intermediate nodes inside the logic cones, along with their mappings that can affect the logic values at the comparison points for the given counterexample

Some equivalence checkers also produce a list of error candidates accompanied by the numeric probability that the failure can be corrected at each of the error candidate locations. Determining the exact location of the error and how to correct it requires the designer's intervention.

Most equivalence checkers provide the following capabilities:

- A graphical user interface (GUI) for easier debugging
- Hooks to a schematic editor for displaying logic cones and navigating through the logic cones in an orderly manner
- Hooks to the source browser to display the source lines that correspond to the signal assignments in the two logic cones

7.3.6.1 Debugging Guidelines

After performing EC on the design, the results can be viewed using the schematic view features, and the errors can be debugged with the generated counterexample. Following are some guidelines that help in debugging.

- **Same inputs**: When reviewing a counterexample, the first thing to ensure is that the two logic cones have the same inputs. If the inputs are not the same, they are the source of the error. If the inputs are the same, the error lies somewhere inside the logic cone. The logic cones must then be compared in detail, starting at the top of each cone. If the logic functions that drive the top-level output are the same, the inputs to those functions in both cones must be examined to determine if and how they correlate. The inputs that correlate must be collapsed and removed from further consideration. These inputs do not require further attention.
- **Inputs not correlated**: Inputs that do not correlate have problems in the sub-cones driving those inputs. However, if the designer believes that the inputs should correlate, the designer can rerun the tool to determine why the functions at those inputs are not equivalent. The tool produces a new counterexample to illustrate why the two inputs are not equivalent. This process can then be repeated until the error is found.
- **Top cone logic**: If the logic functions at the top of the cones are not the same, the designer has to restructure the logic in one of the cones so that they have the same functions at the top of the cone. This makes it easier to correlate the inputs to the top-level functions of the cones. If the restructuring is not obvious or possible, evaluate the functions at the cone outputs in terms of their inputs and determine why they are different.

Figure 7-7 shows the counterexample output from an equivalence checker when the corresponding primary outputs in the two netlist views are not equivalent. The values at the primary inputs corresponding to the counterexample are shown inside the parentheses. Also shown are the logic values at the intermediate nodes. For the same primary input values, the output values of the two netlist views are different.

To diagnose the cause of the different netlist views in the figure, do the following:

1. Review the inputs of the multiplexor (MUX) gates driving the outputs, because the logic gates driving the outputs are the same. Two of the inputs to the MUX gates in the two views are the same, while the third one is different. It turns out that an AND gate drives the differing inputs to the MUX gate in the two views.

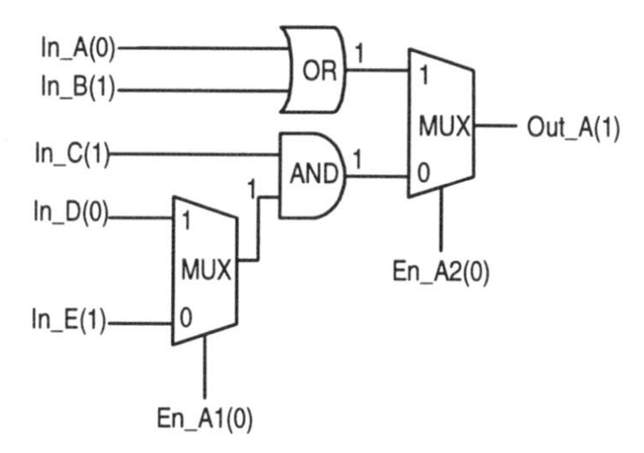

A. Gate-level Netlist View-1.

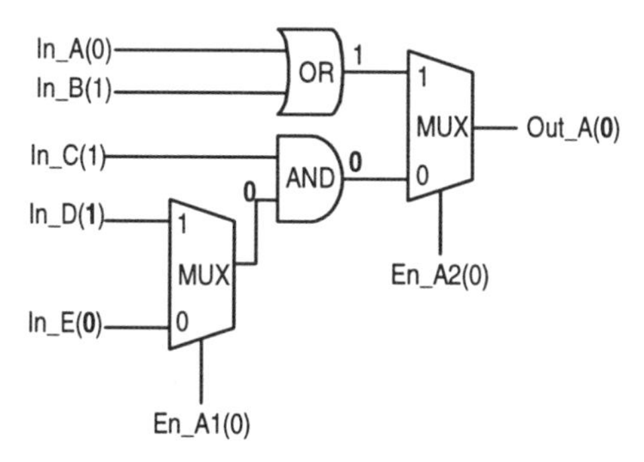

B. Gate-level Netlist View-2.

Figure 7-7. Counterexample Output from an Equivalence Checker

2. Trace the inputs of the AND gate in the two views. One of the inputs of the AND gate is the same in both views, and the other input is different. This input is being driven by a MUX gate in both the views.
3. Review the AND gate inputs and note that the select inputs of the MUXes are the same in the two views, whereas the values at the 1 and 0 inputs are different.
4. Examine the primary inputs D and E. These inputs have been switched in the two views, which is the cause of the inequivalence at the primary outputs.

7.3.7 Performing Equivalence Checking on an Arbiter

This section illustrates implementing an RTL to gate-level netlist EC on the arbiter block used in the Bluetooth SOC design. EC is performed using the Cadence EC solution.

First, the RTL code of the arbiter block is synthesized and a gate-level netlist is generated. The ASPEC sm333s library is used for synthesis. This library contains two files, sm333s.v (standard cells) and prim.v (primitives).The directory structure used is:

- library (top-level)—Contains the above mentioned files
- spec—Contains the RTL code that is used as reference
- impl—Contains the synthesized gate-level netlist
- compare—Comparison is performed in this directory

The following steps are required to run EC for an RTL to gate-level netlist comparison:

1. **Library preparation**: The `libprep` command analyzes, elaborates the library, and creates a hierarchy of library cells, called a reference library, and is used in the design. The libprep script ensures that the library cells are ready for use in EC and are synthesizable.

   Run the following command in the library subdirectory to prepare the ASPEC library for EC:

   ```
   libprep -v "sm33s.v prim.v" -work cellsLib -import
   ```

2. **Design preparation**: Done for both RTL code and gate-level netlists. For the RTL netlist, run the following command.

   ```
   dp -design SpecLib -top asb_arbiter spec/arbiter.v
   ```

   The gate-level netlist library also needs to point to the reference library:

```
dp -design implLib -top asb_arbiter  -reflib cellsLib
impl/arbiter_gate.v
```

The dp script analyzes and elaborates a given Verilog design, thus compiling it into a complete design hierarchy, called a design library, in which all instances in the design are linked together.

3. **Design comparison**: Two generated design views are compared for final EC between two netlists. The following command is used to compare specification and implementation design libraries.

```
heck -spec specLib -ms asb_arbiter -impl implLib -mi
asb_arbiter
```

4. **Report file**: The tool generates a report file for viewing the result information. For this comparison, it reports that both specification and implementation designs are equivalent. Here is an excerpt from the log file for the example:

Elapsed time to parse and compile cycle models: 0.438000 seconds.
The specification has 78 inputs; the implementation has
78 inputs. 78 are common to both.
The specification has 37 outputs; the implementation has
37 outputs. 37 are common to both.
Elapsed time to simplify the netlist: 0.296000 seconds.
Cutting latch boundaries...
Number of spec latches to be refined: 76 out of 76.
Number of impl latches to be refined: 107 out of 107.
:
:
:
Verification succeeds for spec:bd[31] versus impl:bd[31].
Number of outputs checked: 37.
HECK result: DESIGNS EQUAL
Elapsed time to check equivalency: 1.224000 seconds.
Total elapsed time: 2.372000 seconds.

The following report file is generated after running the EC for the arbiter block.

```
REPORT
        Specification          Implementation
        -------------          --------------
DESIGN     specLib.smv           implLib.smv
Top Module ASB_ARBITER           ASB_ARBITER

           Inputs Outputs Latches  Inputs   Outputs Latches
```

```
Total              78    37    76        78           37    107
Common to both      78    37        78           37
Redundant                                                  35


RESULT
EQUIVALENT. Subject to the following conditions the spec
module can be replaced by the impl module.

CONDITIONS

   Initial conditions:
Initial conditions:
        group #1. impl:bwait_r_reg_TRI.Q
          impl:blast_r_reg_TRI.Q
          impl:berro_r_reg_TRI.Q
        .................

WARNINGS
Some implementation latches are redundant, assuming a
common initial state; if the designs are shown to be
equivalent, it is in the context of
this common initial state assumption.
ERRORS
None.
DETAILS
Functional latch mapping correspondence in file rundir/
function_leq.map
List of functionally mapped latches in file rundir/
latchlist.out

Outputs verified: 37
      spec:agnt0    vs impl:agnt0
      spec:agnt1    vs impl:agnt1
....
```

7.4 Static Timing Verification

STV is a critical element of any design verification methodology. Each storage element and latch in a design have timing requirements that have to be met. These include setup, hold, and various delay timings. Timing verification is challenging for a complex design, since each input can have multiple sources and the timing can vary depending on the circuit operating condition.

In a chip design, the netlist of the design goes through transformation after synthesis, scan insertion, clock-tree generation, floorplanning, block placement, and complete chip routing. STV is performed after each of these steps to ensure that the design timing requirements are met. STV performs an exhaustive analysis of all the paths in a design and does not require test vectors as in simulation.

There are two types of timing analysis: dynamic and static. Dynamic timing analysis uses simulation vectors to verify that the circuit or design outputs accurate results from a given input without any timing violations. The problem with dynamic timing analysis is that the simulation vectors cannot guarantee 100 percent coverage.

Static timing analysis checks all the paths in the circuit or design, including the false paths. The static timing analysis can guarantee 100% coverage. The static timing analysis does not verify the circuit functionality. Static timing analysis is widely used by many design houses.

To perform STV, the following items should be defined:

- Model of the design
- All points that require timing check, such as inputs, outputs, storage elements, latches, and gates
- All timing constraints, such as setup, hold, and delay
- Boundary conditions, such as input arrival-time, input slew-rate, output required-arrival time constraints, and output capacitance loading constraints
- Clock details

7.4.1 Selecting an STV Solution

Following are some of the areas to take into consideration when selecting an STV solution.

- **Input**: Check the input design formats that the solution accepts. The solutions currently available accept netlist in Verilog, VHDL, EDIF, and SPICE formats. Some tools also accept the physical netlist in the DEF format, enabling a seamless embedding of the tool in both logical and physical phases of the chip design.

- **Accuracy**: This is a crucial factor. The accuracy obtained depends on the accuracy of the models used to represent the design timings.
- **Netlist**: Should support flat or hierarchical netlists analysis capabilities.
- **Design capacity**: What design complexity the solution can handle. For example, some tools can verify a 1 million gate netlist in less than 10 minutes on a workstation.
- **Run time**: This a critical factor, since STV is performed many times in the chip design methodology flow. A longer run time can significantly impact the overall schedule.
- **SDF**: Should accept back-annotated resistance (R) and capacitance (C) or delays in SDF or SPF formats.
- **Analysis**: Should be user friendly, such as a GUI, TCL, and text-based analysis commands to analyze chip timings.
- **Support**: Should support handling gate array, standard cell, and custom design styles.
- **Host requirements**: What are the system memory, disk space, computation time capabilities.

7.4.2 STV Methodology

The following steps are involved in STV, as shown in Figure 7-8.

1. **Inputs**: The STV tool requires the gate-level netlist of the design, the cell library file, and timing constraint files.
2. **Read in files**: All files inputted to the tool are read in to perform the timing analysis.
3. **Clock**: The clocks and cycle time are specified as per the design requirements.
4. **Run**: Verification is run on the input files and the clocks as per the design requirements. The results are generated in the form of reports.
5. **Analyze**: The reports generated by the STV tool are analyzed to see whether the design meets the timing requirements. Timing violations are identified and handed off to the design team to resolve.

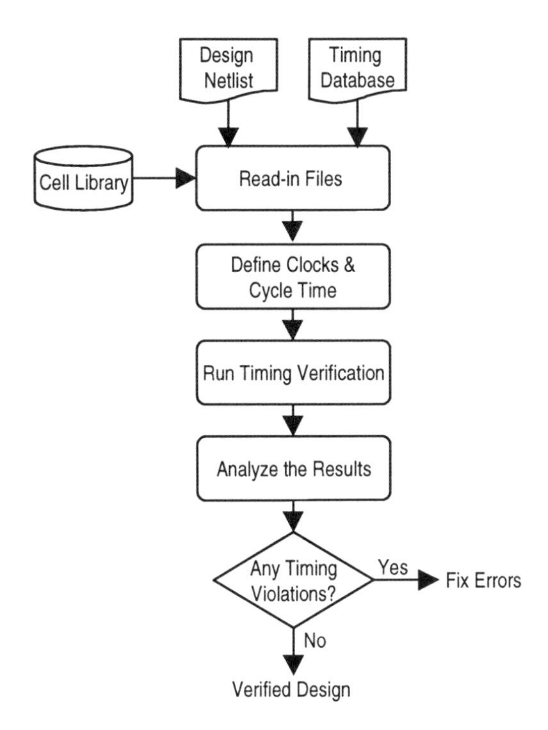

Figure 7-8. STV Methodology Flow

Some of the guidelines for performing timing verification are as follows.

- **Information**: Gather and study all the information required, such as models, timing assertions, etc.
- **Define clocks**: Clock signals need to be defined accurately, since the timing verification tool uses the clock information for generating the arrival and the required-arrival times at the storage elements and latches.
- **Clock and data nets**: Do not mix the clock with data signals on the same net. The clocks should exist in the clock-tree only.
- **Reference time**: Specify a common reference time for the clock signal, and all input-arrival and output required-arrival times.
- **Signal-arrival time**: Check whether minimum or early-arrival time values are maximum or late-time values.

- **Avoid "don't care" directives**: If "don't care" directives are used, the timing analyzer ignores parts of the logic.
- **Results**: After running the timing verification on a design, check that the clock signals are reaching the inputs of latches or flip-flops at expected times. If the clocks are not reaching as expected, check for the parameters that are set in the tool and rerun the verification.

7.4.3 Performing STV on an Arbiter

This section illustrates implementing STV on the arbiter block used in the Bluetooth SOC design, using the Cadence static timing analysis solution to trace and analyze all the paths for the gate-level netlist.

The gate-level netlist of the arbiter block is shown in Example 7-1. This is generated after synthesizing the RTL code. The library targeted for synthesis is ASPEC sm333s.

Example 7-1: Gate-Level Netlist of Arbiter Block

```
module asb_arbiter(Areq0, Areq1, Dsel, BWRITE, BSIZE,
BWAIT, BLAST, BERROR, BD, BnRes, BCLK, Agnt0, Agnt1);

        input Areq0;
        input Areq1;
        input Dsel;
        input BWRITE;
        input [1:0] BSIZE;
        inout BWAIT;
        inout BLAST;
        inout BERROR;
        inout [31:0] BD;
        input BnRes;
        input BCLK;
        output Agnt0;
        output Agnt1;

ssao21          i_13(.A(\current_state[0]          ),
.B(\current_state[1] ), .C(Areq0), .Y(n_20));
ssnd2 i_2409(.A(n_16), .B(n_18), .Y(n_19));
```

```
ssnd5  i_8(.A(Areq1),  .B(Areq0),  .C(n_45),  .D(n_47),
.E(n_46), .Y(n_18));
ssoa21 i_7(.A(n_1447), .B(n_14), .C(Areq0), .Y(n_16));
ssao21 i_6(.A(\current_state[1] ), .B(\current_state[0]
), .C(Areq1), .Y(n_14));
ssnd2a i_17(.A(n_30), .B(n_1441), .Y(n_23));
ssnd2a i_2517(.A(n_1423), .B(n_23), .Y(n_24));
ssnr2 i_2581(.A(n_1423), .B(n_1441), .Y(n_25));
ssoa21   i_2589(.A(\current_state[0]   ),   .B(n_49),
.C(n_23), .Y(n_26));
ssnd3 i_1(.A(Areq0), .B(n_46), .C(\RID[0] ), .Y(n_30));
ssao21 i_14(.A(n_30), .B(n_47), .C(\current_state[1] ),
.Y(n_31));
ssoa21    i_2415(.A(n_20),    .B(n_31),    .C(Areq1),
.Y(\next_state[1] ));
ssad4 i_1706(.A(Areq1), .B(n_1447), .C(n_45), .D(n_46),
.Y(n_1423));
ssld1q RID_reg_0(.D(n_1423), .G(n_24), .Q(\RID[0] ));
ssld1q RID_reg_1(.D(1'b0), .G(n_24), .Q(\RID[1] ));
ssfd6q    current_state_reg_0(.D(n_19),    .CKN(BCLK),
.RN(BnRes), .Q(\current_state[0]));
ssnr2    i_2455(.A(\current_state[1]    ),    .B(n_47),
.Y(n_1441));
ssfd6q    current_state_reg_1(.D(n_48),    .CKN(BCLK),
.RN(BnRes), .Q(\current_state[1] ));
ssnr2    i_2476(.A(\current_state[0]    ),    .B(n_49),
.Y(n_1447));
ssld5q Agnt0_reg(.D(n_1441), .GN(n_25), .Q(Agnt0));
ssld1q Agnt1_reg(.D(n_1447), .G(n_26), .Q(Agnt1));
ssiv i_74(.A(\RID[0] ), .Y(n_45));
ssiv i_75(.A(\RID[1] ), .Y(n_46));
ssiv i_76(.A(\current_state[0] ), .Y(n_47));
ssiv i_77(.A(\next_state[1] ), .Y(n_48));
ssiv i_78(.A(\current_state[1] ), .Y(n_49));
endmodule
```

7.4.3.1 Running Timing Analysis on the Arbiter Block

The gate-level analysis of the cell-based netlist, in which every cell is characterized by a timing model, is shown in Example 7-2.

Example 7-2. Excerpts from Timing Analysis Run

```
cmd> include arbiter.cmd
arbiter.cmd> logfile arbiter.log
arbiter.cmd> readtechnology ./std-cell.tech
arbiter.cmd> readctlf ./sm333s.ctlf
arbiter.cmd> readverilog ./arbiter_gate.v
arbiter.cmd> toplevelcell asb_arbiter
Network asb_arbiter has 26 devices and 71 nodes
arbiter.cmd>  buildtimingmodel  -format  tlf  -header  -
input_slew 0.01,0.02,0.04,0.0
8,0.16,0.32,0.64,1.28,2.56,5.12                 -load_cap
0.012,0.024,0.048,0.096,0.18,0.36,0.72
,1.42,2.84,6.68 arbiter.tlf
 1:    0.27ns Path from BCLK v to Agnt1 v
 2:    0.26ns Path from BCLK v to Agnt1 ^
 3:    0.25ns Path from BnRes v to Agnt1 v
 4:    0.24ns Path from BCLK v to Agnt0 v
 5:    0.24ns Path from BCLK v to Agnt0 ^
 6:    0.24ns Path from BnRes v to Agnt1 ^
  7:    0.22ns Path from BnRes v to Agnt0 v
 8:    0.22ns Path from BnRes v to Agnt0 ^
 9:    0.17ns Path from Areq1 ^ to Agnt1 v
10:    0.17ns Path from Areq0 ^ to Agnt0 v
```

The `include arbiter.cmd` command performs alias setting and specifies the input files to be read. The `readtechnology` command reads the standard cell technology, specifies power and ground node names, the logic threshold, the default rise and fall times on the input nodes, and the data to estimate stray capacitances. `readctlf` reads in the compiled timing library for the timing models used by the arbiter gate netlist. The `readverilog` command reads in the Verilog code. `toplevelcell` sets the top-level to asb_arbiter. The `buildtimingmodel` command builds the TLF timing model for the circuit, allowing the circuit to be replaced with a single-timing model in the next higher level of the hierarchy.

The following command identifies the longest paths triggered by the falling edge of the clock:

```
cmd> findpathsfrom BCLK v
```

The result of running the above command is as follows. For each path, it gives the delay and end of node.

```
 1:    0.28ns Path to n_19 ^
 2:    0.27ns Path to n_48 ^
 3:    0.27ns Path to Agnt1 v
 4:    0.26ns Path to Agnt1 ^
 5:    0.25ns Path to n_19 v
 6:    0.24ns Path to Agnt0 v
 7:    0.24ns Path to Agnt0 ^
 8:    0.23ns Path to n_48 v
 9:    0.22ns Path to n_26 v
10:    0.20ns Path to n_1423 ^
```

The delay between two nodes can be calculated with the following command:

```
cmd> showeqnsbetween n_49 v n_1447 ^
```

This results in the following:

```
n_49 v -> n_1447 ^ 0.03ns
Delay = TLFdelay(i_2476 B v -> Y ^)
        Cout = 0.05pf
        Tin = 0.01ns
```

The delay on node n_49 can be determined using the following command:

```
cmd> describenode n_49
```

The result of running this command is:

```
Node n_49
 Capacitance: 0.02pf
  0.02pf pin
2 input pins
  i_2476/B 0.01pf
  i_2589/B 0.01pf
1 output pins
  i_78/Y 0.00pf
```

The clock period for 40 ns is specified as follows:

```
cmd> waveform -name ideal_clk -period 40 -rise_first
     0.0 20
cmd> clockwaveform BCLK ideal_clk
```

The `waveform` command defines the clock waveform. `clockwaveform` associates a waveform with the clock root.

The timing of the design can be verified by issuing the `timingverify` command:

```
cmd> timingverify
```

This command summarizes the register and latch setup and hold timing checks, as follows:

```
1:Setup constraint slack  39.94ns Agnt0_reg (ssld5q D
^ -> GN ^)
2:Setup constraint slack  39.96ns Agnt1_reg (ssld1q D
^ -> G v)
3: Setup constraint slack 39.97ns Agnt0_reg (ssld5q D
v -> GN ^)
4: Setup constraint slack 39.99ns Agnt1_reg (ssld1q D
v -> G v)
5: Hold  constraint violation 0.11ns Agnt0_reg
(ssld5q D v -> GN ^)
6: Hold  constraint violatio  0.10ns Agnt1_reg
(ssld1q D v -> G v)
7: Hold  constraint violation 0.09ns Agnt0_reg
(ssld5q D ^ -> GN ^)
8: Width constraint violation 0.19ns RID_reg_1
(ssld1q G high)
9: Width constraint violation 0.19ns RID_reg_0
(ssld1q G high)
10: Width constraint slack    19.92ns
current_state_reg_1 (ssfd6q CKN low)
```

For the details of one of the above timing checks, use the `showpossibility` command:

```
cmd> showpossibility
```

Running this command results in the following. The Delay column shows the cumulative delay of the node. Delta is the delay from node to next node. Node is a node on the delay path with its edge, ^ for rise or v for fall. Device is the device on the path triggered by the node. Cell is the type of device.

```
Possibility 1:
Setup  constraint  slack  39.72ns  current_state_reg_0
(ssfd6q D ^ -> CKN v)
  Clk edge: BCLK v at 20.00ns + Tcycle = 60.00ns
  Setup time: 0.00ns
  Data edge: BCLK v -> n_19 ^ at 20.28ns
  Required cycle time: 0.28ns (1.00 cycle path)
    Delay   Delta  Node         Device              Cell
    -----   -----  ----         ------              ----
* 20.00ns 0.06ns BCLK v    current_state_reg_1   ssfd6q
  20.06ns 0.00ns  current_state\[1\] ^ i_78       ssiv
  20.06ns 0.03ns n_49 v         i_2476              ssnr2
* 20.10ns 0.04ns n_1447 ^       i_1706              ssad4
  20.14ns 0.02ns n_1423 ^       RID_reg_0           ssld1q
  20.15ns 0.02ns RID\[0\] ^     i_1                 ssnd3
  20.17ns 0.01ns n_30 v         i_17                ssnd2a
  20.19ns 0.01ns n_23 v         i_2517              ssnd2a
* 20.20ns 0.04ns n_24 ^         RID_reg_1           ssld1q
  20.23ns 0.01ns RID\[1\] v     i_75                ssiv
  20.24ns 0.03ns n_46 ^         i_8                 ssnd5
  20.27ns 0.01ns n_18 v         i_2409              ssnd2
```

Summary

Equivalence checking and static timing verification are efficient verifications techniques for SOC designs. Although these tools may take some time to become familiar with, verification time is ultimately reduced, helping to get products to market quicker.

References

1. An integrated formal verification solution for DSM sign-off market trends, www.cadence.com.

2. Clayton Shawn, Sweeney John, ... A set of formal applications, Integrated System Design, November 1996.

3. Static functional verification with Solidify, A whitepaper, www.hdac.com.

4. FormalPro data sheet, www.mentor.com.

5. Schroeder Scott. Turning to formal verification, Integrated System Design, September 1997.

6. Tuxedo-LEC, Logic equivalence checker data sheet, www.verplex.com.

7. Cadence equivalence checker data sheet and users manual, www.cadence.com.

8. Parash Avi, Formal verification of an MPEG decoder chip, Integrated System Design, August 2000.

9. Formality, Formal verification equivalence checker data sheet, www.synopsys.com.

10. Schulz Steven E. Focus report: Timing analysis, Integrated System Design, August 2000.

11. Pan Jengwei, Biro Larry, ... Timing verification on a 1.2M-device full-custom CMOS design, www.sigda.acm.org.

12. Cadence Pearl® timing analyzer data sheet and users manual, www.cadence.com.

13. Synopsys Prime-Time data sheet, www.synopsys.com.

14. Granese Paul. Using timing analysis for ASIC sign-off, Integrated System Design, May 1995.

15. Bassak Gil. Focus report: Timing and power analysis, Integrated System Design, August 1998.

16. Schulz Steve, Deep-submicron timing closure, Integrated System Design, June 2000.

17. Bronnenberg Dean. Static timing analysis increases ASIC performance, Integrated System Design, June 1999.

18. Huang Shi Yu, Cheng Kwang Ting. Formal Equivalence Checking and Design Debugging, Kluwer Academic Publishers, 1998.

CHAPTER 8

Physical Verification and Design Sign-off

In the past, chip physical design processes required only design, electrical rules checking, and layout versus schematic check in physical verification. Now, in system-on-a-chip (SOC) designs with deep sub-micron (DSM) geometries, the traditional rules checking is not sufficient. As the chip design processes are shrinking, the interconnect delays are dominating the gate delays. The average interconnect line is lengthening as the processes shrink. The latest chip designs operate at higher clock speeds and are created using many blocks, with many lengthy interconnections running between the blocks. The interconnect delay will continue to dominate, since most designs will be based on reusable design blocks that are interconnected for the intended functionality. In addition, the latest designs are based on embedded cores connected on buses that have lengthy lines. The physical verification tools and methodologies for SOC need to address timing, power, signal integrity, electromagnetic interference, metal migration, and thermal effects as well as rules checking. This chapter briefly illustrates the following:

- Design checks
- Physical effects and analysis
- Design sign-off

The detailed methodology for each of these physical verification steps is beyond the scope of this book. However, the physical verification terminology of the processes used for the current DSM chips is covered.

8.1 Design Checks

Design checks are performed on the design after the placement and routing process is completed to ensure that the design is error free and ready for fabrication.

Design rules checking verifies that no errors occurred in the placement of the logic cells and the routing process. In design rules checking, the focus is to find violations related to spacing, connections, vias, and so on. The chip manufacturer also performs design rules checking when accepting the design for fabrication.

Electrical rules verifies that no errors occurred in the placement of the logic cells and the routing process. In electrical rules checking, the focus is to find short circuits, open circuits, and floating nodes.

Layout versus schematic is performed on the design after routing to ensure that the final physical layout consistent to the input netlist. To perform layout versus schematic, an electrical schematic is extracted from the physical layout and compared with the input netlist.

8.2 Physical Effects Analysis

In DSM designs, many electrical issues must be analyzed to ensure correct operation. The electrical issues include timing, signal integrity, crosstalk, IR drop, electromigration, and power analysis.

8.2.1 Parasitic Extraction

In DSM designs, the interconnect parasitics must be accurate and should be considered early in the chip design process. This requires accurate estimation of the parasitics in the pre-layout stage, and accurate extraction and analysis in the post-layout stage of a chip design.

The three methodologies used for parasitics extraction are:

- **2-D**: This is a simplest method of extraction. In this method, it is assumed that the geometries modeled are uniform in signal propagation direction, and all three dimensional details are ignored. When it is used for DSM designs, it gives inaccurate results because the 3-D field effects, such as capacitive coupling between geometries and non-orthogonal cross sections, cannot be modeled accurately.

- **2.5-D** or **Quasi 3-D**: This method can model 3-D structures more accurately than 2-D.
- **3-D**: This method uses full 3-D field solution for parasitics capacitance extraction. It is the most accurate method. For large designs, it requires a long execution time for detailed extraction.

The 2-D method is used for simple and lumped capacitance values for all nets. Using the extracted data, the critical nets are identified based on the length, drive strength, and loading. The critical nets are further extracted using the 3-D method for better accuracy.

8.2.2 Inductance Effects

In DSM chips, inductance effects in on-chip interconnects have become significant for specific cases, such as clock distributions and other long, low-resistance on-chip interconnects optimized for high performance. The phenomenon of inductive coupling is negligible at short trace interconnects, since the signal edge rate is long compared to the flight time of the signal. The inductive coupling effect becomes significant for long interconnects and for very fast signal edge rates. Accurate on-chip inductance extraction and simulation are much more difficult than capacitance extraction.

8.2.3 Signal Integrity

Signal integrity is the ability of a signal to generate correct responses in a circuit. A signal with signal integrity has digital levels at appropriate and required voltage levels at required instants of time. A lack of signal integrity leads to erroneous data and can result in a faulty prototype or production chip. Maintaining signal integrity in high-speed and DSM designs is a very challenging task. The sources of interference need to be fixed early in the design cycle to eliminate problems. This is only possible when suitable tools and methodologies are adopted or created. The signal integrity analysis needs to be incorporated into every stage of the design process or methodology, as shown in Figure 8-1 for both block-level and chip-level design. The analysis performed at the block level is fed as input to the chip-level steps for overall chip-design analysis.

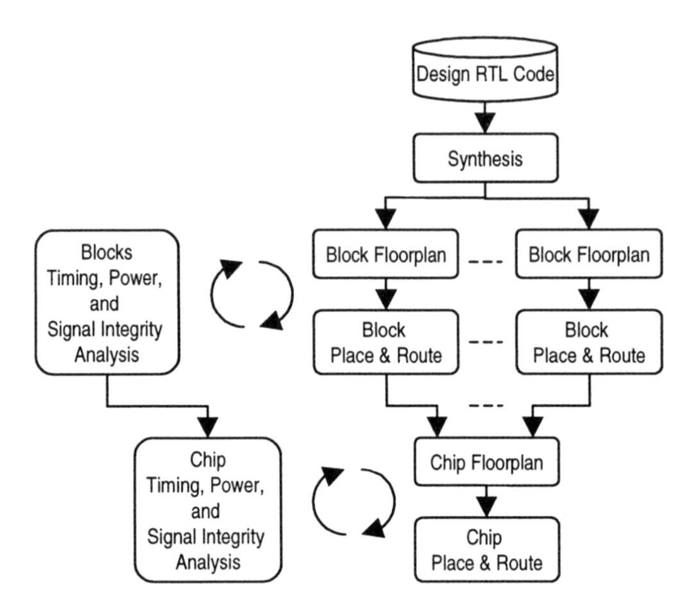

Figure 8-1. Signal Integrity Analysis Steps

8.2.3.1 Crosstalk

Crosstalk is the interaction between signals on two different electrical nets within an integrated chip (IC). The net creating the crosstalk is called an aggressor, and the net receiving it is called a victim. A net may be both an aggressor and a victim. Figure 8-2 shows the effect of an aggressor net on the victim net.

The victim experiences a delay when both aggressor and victim switch simultaneously. When the aggressor only switches and the victim is quiescent, there is a crosstalk effect on the victim. The magnitude of the effect of the aggressor net on the victim net depends on the coupled capacitance and resistance of the driver. To limit the crosstalk within specified margins, a buffer (repeater) cells are is used between the transmitting point and the receiving point. The crosstalk can also be reduced by increasing the spacing between the wires (this reduces the coupled capacitance) and increasing the width of the wires (this reduces the resistance of the driver), but over-designing increases the chip size and decreases the yield.

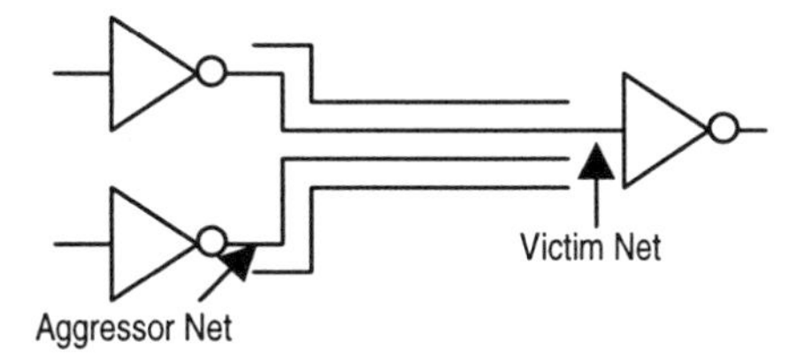

A. Effect of Aggressor net on Victim Net.

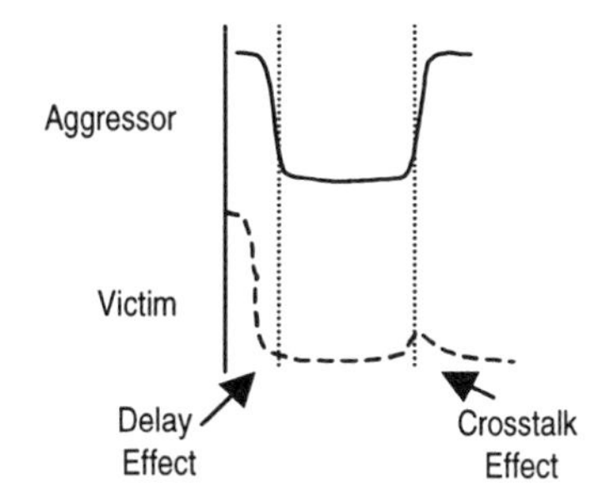

B. Waveforms on Aggressor and Victim Nets.

Figure 8-2. Signal Integrity

8.2.3.2 IR Drop

The voltage drop at the gates or nodes due to the resistance of the power supply lines is called IR drop. In low voltage DSM designs, the higher switching frequencies and thinner wires cause functional failures because of excessive IR drop. IR drop weakens the driving capability of the gates, lowers the noise margin, and increases the overall delay. Typically the gate delay increases by 15 percent for a 5 percent IR drop in the power supply. This requires the IR drop or the power to be estimated at the pre-layout stage and analyzed after post-layout stage in the chip design process.

Today, most tools focus on the post-layout, transistor-level verification of a power distribution. The transistor-level power estimation is accurate, but requires a very long execution time. If any problems arise due to IR drop at the post-layout stage, it is very costly to fix them at this late stage, therefore the estimation and analysis of power distribution should be done early in the design cycle.

Dynamic gate-level power estimation is promising for the full chip, because the transistor-level power estimation requires a long execution time and power estimation at register-transfer level (RTL) is not accurate. Dynamic gate-level power estimation is less accurate than transistor-level, but takes considerably less execution time. Gate-level power estimation can be used for both pre-layout and post-layout stages in the chip design. As shown in Figure 8-1, power estimation is done at the floorplan stage of every block. The power estimation data at the block level is fed as input to the chip-level power estimation. Similarly, power analysis is done after the place and route to check for power violations.

The most critical and fastest nets in any chip are usually the clocks. The performance of the chip depends on the quality of the clock and the clock distribution circuit strategy used. Clocks in DSM chips have very high frequencies, and they need to drive a large number of gates or loads. This can lead to significant IR drop and degrade the quality of the clocks because of a higher load on the power supply distribution network. To ensure that the chip has high-quality clock signals, it is essential to verify the clocks accurately.

8.2.4 Electromigration Effects

In DSM designs, electromigration is the mass transport in the metal lines, under high current and high temperatures. Over the lifetime of the chip, the electrons flow through the wires and collide with metal atoms, producing a force that causes the wires to break. This problem is due to the high current densities and high frequencies going through the long, very thin metal wires. Electromigration occurs on power grids and signal lines. The effect can be eliminated by using the appropriate interconnect wire sizing.

8.2.5 Subwavelength Challenges

In DSM designs below a 0.25 micron process, the chip feature sizes are smaller than the wavelength of exposure light that is used in optical equipment for manufacturing the chip. This requires significant changes in the methodologies used for chip manufacturing. In recent years, two software technologies, phase shift mask (PSM) and optical proximity correction (OPC), have emerged to address these manufacturing issues. PSM enables smaller geometries, and OPC addresses fixing subwavelength distortion.

8.2.5.1 Phase Shift Mask

PSM employs phase and interference to provide sharper images for improving optical resolution. The phase shifters are placed near each other on the mask, 180 degrees apart, to create interference patterns. The patterns can generate features smaller than the conventional optical resolution limit of lithography steppers.

8.2.5.2 Optical Proximity Correction

OPC enables smaller features in a chip to be produced using given equipment by enhancing the lithographic resolution. OPC resolves distortions that result from optical proximity, diffusion, and loading effects of resist and etch processing. It does this by adding features to the chip layout at the mask level. The corrections that are addressed include enhancing outside corners of the design, trimming the inside corners of the geometries to prevent excessive rounding or line shortening, and preventing shorts and opens.

Using OPC in the manufacturing process influences the design, mask making, and fabrication steps. This requires that manufacturers fabricate special test chips to generate OPC models. Designers then need to use the OPC models in the design and create the optical proximity-corrected mask layers.

8.2.5.3 Verification after PSM and OPC

Design checks that are used for physical verification, such as design rules checking, layout versus schematic, and parasitic extraction, use the original physical layout as input. These steps use a rules-based process description that assumes that the physical layout represents the final silicon. With PSM and OPC, the original layout is modified, and it is no longer an accurate representation of the final silicon pattern. Therefore, OPC and PSM issues need to be addressed early in the design cycle to perform estimation, extraction, and analysis.

8.2.6 Process Antenna Effect

Modern chip technologies use plasma-based manufacturing processes for etching, ashing, and oxide depositions to achieve the fine feature size of chips. Unfortunately, these processes cause antenna effects, also known as plasma-induced, gate-oxide damage. The plasma etchers can induce a voltage into isolated leads, stressing the thin gate-oxides. The metal or polysilicon leads act like an antenna and collect the charges. The accumulated charges can result in oxide breakdown. The process antenna effect directly impacts the yield and reliability of the chips. This is a major manufacturing issue and should be addressed by the design methodology and process improvements.

8.3 Design Sign-off

Sign-off is the final step in the design process. The chip design is ready to be taped-out for fabrication. At this point, everything about the chip design is known to the design team. This is a very critical step, since the chip design is committed for the physical prototype or production, and no corrections can be made. The design team needs to be confident that the design works as per the intended functionality and performance. Figure 8-3 shows the design sign-off steps.

The following is a list of items that need to be checked for the final design sign-off.

- **Design**: Is the design synchronous.
- **Asynchronous signals**: Are all asynchronous set and reset signals resynchronized with the clock.
- **Memory**: Are the memory read/write signals synchronized at the memory boundary. Check for memory block design, guard-band, placement, and the corresponding design rules.
- **AMS**: Do the analog/mixed signal block design, guard band, and placement correspond to the design rules.
- **Buses**: Does the bus protocol meet the intended requirements.
- **Latches**: There should be no latches in the control and random logic.
- **Combinational paths**: There should be no direct combinational paths from inputs to outputs.
- **Multicycle paths**: There should be no multicycle paths.
- **Feedback loops**: There should be no combinational feedback loops.
- **Registers**: Every block's inputs/outputs should be registered.
- **Tri-states**: There should be no tri-stated logic internal to the design.
- **Speed**: Does the design meet the clock speed requirements.

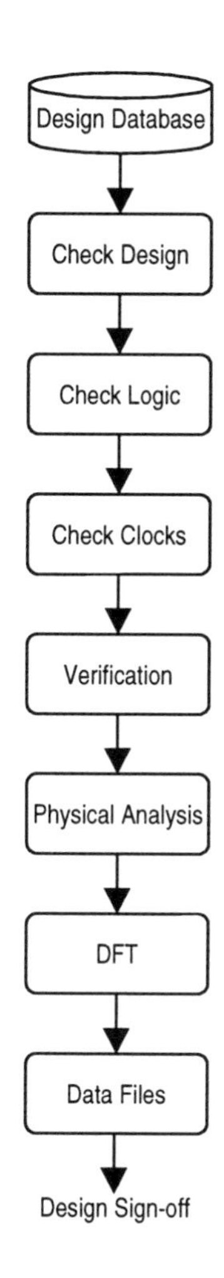

Figure 8-3. Design Sign-off Process

- **Buffers**: Are the clocks buffered at the boundary of every block.
- **Multiple clocks**: Are the multiple clock domains isolated.
- **Load balancing**: Is the clock tree or network load-balanced.
- **Skew**: Is the clock skew within the specified margin limits.

- **RTL code coverage**: Is the design's RTL code fully covered, that is, are all the paths in the code are tested.
- **Transaction function coverage**: Does the design pass all the transaction function coverages that are defined if transaction-based verification is used.
- **Simulation**: Does the design meet the intended functionality when performing functional simulation and gate-level simulation.
- **Formal equivalence checking**: Is the final netlist equivalent to the input design using formal equivalence checking.
- **Timing**: Is the timing analysis done, and does the design meet the required timing specifications.
- **Standard delay format**: Is the back annotation done using the design's standard delay format file.
- **Rules**: There should be no violations in the design and electrical rules checks, and the layout versus schematic.
- **Crosstalk**: Is the crosstalk within the specified margin.
- **IR drop**: Is the IR drop within the specified margin.
- **Coverage**: Was fault coverage done on the design.
- **Simulation**: Was fault grading and simulation done.
- **Scan rules**: Was scan rules checking done.
- **Data files**: In this step, the files and scripts that are required for the fabrication, such as synthesis, timing, test, and design shell, are checked for correctness.

Summary

Because of their size and complexity, SOC designs require that new physical verification tools and methodologies be adopted. It is critical that all the design issues be checked prior to sign-off.

References

1. Chen Howard H, Ling David D. Power supply noise analysis methodology for deep-submicron VSLI chip design, Design Automation Conference 1997.

2. NS Nagaraj, Can Frank, ... A practical approach to static signal electromigration analysis, Design automation conference '98.

3. Li Tong, Kang Sung-Mo. Layout extraction and verification methodology for CMOS I/O circuits, Design Automation Conference 1998.

4. Restle Phillip, Ruehli Albert, Walker Steven G. Dealing with inductance in high-speed chip design, Design Automation Conference 1999.

5. Beattie Michael W, Pileggi Lawrence T. IC analysis including extracted inductance models, Design Automation Conference 1999.

6. Morton Shannon V. On-chip inductance issues in multiconductor systems, Design Automation Conference 1999.

7. Cong Jason, He Lei, ... Analysis and justification of a simple, practical 2 1/2-D capacitance extraction methodology, Design Automation Conference 1997.

8. Kahng Andrew B, Pati Y C. Subwavelength lithography and its potential impact on design and EDA, Design Automation Conference 1999.

9. Smith Wayne, Trybula Walt. Photomasks for advanced lithography, IEEE Int'l Electronics Manufacturing Technology Symposium, 1997.

10. Jain Nirmal, Silvestro John, ... SI issues associated with high speed packages, IEEE Electronic Packaging Technology Conference, 1997.

11. Gal Laszlo. On-chip cross talk - the new signal integrity challenge, IEEE Custom Integrated Circuits Conference, 1995.

12. NS Nagaraj, Can Frank, ... A practical approach to crosstalk noise verification of static CMOS designs, International Conference on VLSI Design, January 2000.

13. Maniwa R T. Focus report: Signal integrity tools, Integrated System Design, July 1996.

14. Green Lynne. Addressing the effects of signal integrity in deep-submicron design, Integrated System Design, July 1998.

15. Gupta Rohini, Tauke John. Addressing signal integrity in deep-submicron SOC designs, Integrated System Design, April 2000.

16. Maniwa Tets. Focus report: Physical verification, Integrated System Design, January 2000.

17. Hussain Zakir Syed, Rochel Steffen, ... Clock verification in the presence of IR-drop in the power distribution network, IEEE Custom Integrated Circuits Conference, 1999.

18. Saleh R, Overhauser D, Taylor S. Full-chip verification of UDSM designs, ICCAD 1998.

19. Parasitic extraction for deep submicron and ultra-deep submicron designs, A technical paper, www.simplex.com.

20. Calibre data sheets, www.mentor.com.

21. Saal Frederick. A case for signal integrity verification, EE Times, January 17, 2000.

22. Powell Jon. Solving signal-integrity problems in high-speed digital systems, EDN Access, www.ednmag.com.

23. Cadence Dracula®, Diva, Vampire®, Assura™ SI users manuals, www.cadence.com.

Glossary

ADC, A/D—Analog-to-Digital Converter.

AHDL—A Hardware Description Language, such as Verilog-A, SpectreHDL, or VHDL-A, used to describe analog designs.

AMBA—Advanced Microcontroller Bus Architecture. An on-chip bus released by advanced risc machines (ARM).

AMS—Analog/Mixed Signal. The combination of analog and digital technology on the same integrated circuit (IC).

APB—Advanced Peripheral Bus. An on-chip bus released by advanced risc machines (ARM).

ARM7TDMI—A family of RISC processors from Advanced Risc Machines (ARM). Refer to www.arm.com for more details.

ASB—Advanced System Bus. An on-chip bus released by advanced risc machines (ARM).

ASIC—Application Specific Integrated Circuit.

ATE—Automatic Test Equipment.

ATM—Automatic Transfer Mode.

ATPG—Automatic Test Pattern Generator.

BDM—Background Debug Mode. An on-chip debug mode available in Motorola microcontrollers.

BFM—Bus Function Model.

BIC—Bus Interconnection Device.

BLC—Bluetooth Link Controller.

Bluetooth—An open protocol standard specification for short-range wireless connectivity. Refer to www.bluetooth.com for more details.

BSP—Board Support Package.

C, C++—Programming languages used for software development.

CAD—Computer Aided Design.

CAS—Cycle Accurate Simulator.

CBS—Cycle-Based Simulation.

Certify—An FPGA synthesis, partitioning, and configuration tool available from Synplicity.

Chip—A single piece of silicon on which a specific semiconductor circuit has been fabricated.

ConCentric—A system-design tool available from Synopsys.

Core—A complex, pre-designed function to be integrated onto a larger chip, such as PCI, MPEG and DSP functions, microprocessors, microcontrollers, and so on. The core is also called a macro, block, module, or virtual component.

COSSAP—A system-design tool available from Synopsys.

Coverscan—A code coverage tool available from Cadence.

DAC, D/A —Digital-to-Analog Converter.

DEF—Design Exchange Format. A Cadence format used to describe physical design information. Includes the netlist and circuit layout.

Design flow—The process of a chip design from concept to production.

Design house—A company specializing in designing ICs, but has no in-house manufacturing and does not sell its designs on the open market.

Design reuse—The ability to reuse previously designed building blocks or cores on a chip for a new design as a means of meeting time-to-market goals.

Design rules—Rules constraining IC topology to assure fabrication process compatibility.

DFT—Design For Test. Refers to specific activities in the chip design process that provide controllability and observability to determine the quality of the product.

DMA—Direct Memory Access.

DRAM—Dynamic Random Access Memory.

DRC—Design Rules Check

DSM—Deep Sub-Micron.

DSP—Digital Signal Processor. A high-speed, general-purpose arithmetic unit used for performing complex mathematical operations.

DUT/DUV—Design Under Test/Design Under Verification.

EBS—Event-Based Simulation.

EC—Formal Equivalence Checking.

ECO—Engineering Change Order.

EDA—Electronic Design Automation. Describes a group of CAD tools used in the design and simulation of electronic circuits. EDA tools allow designers to describe and test the performance of circuits before they are implemented in silicon. The

EDA suppliers include Cadence, Synopsys, Mentor, and a host of smaller vendors. Refer to www.edacafe.com for more details on EDA companies and the products they offer.

EDIF—Electronic Design Interchange Format.

EPROM—Erasable-Programmable Read-Only Memory.

ERC—Electrical Rules Check.

Equivalence Checker—A formal equivalence checking tool available from Cadence.

ESW—Embedded Software.

Fault Coverage—A measure that defines the percentage of success a test set has in finding simulated stuck-at-0 or stuck-at-1 faults for a list of nodes in a given design.

FFT—Fast Fourier Transform.

FIFO—First In First Out.

Firm Core—IP building block that lies between hard and soft IP. Usually these are soft cores that have been implemented to fully placed netlists.

FormalCheck—Model checking tool available from Cadence.

FPGA—Field Programmable Gate Array. An IC incorporated with an array of programmable logic gates that are not pre-connected, and the connections are programmed by the user.

Foundry—Semiconductor company that fabricates silicon chips.

FSM—Finite State Machine.

Gate—Basic circuit that produces an output when certain input conditions are satisfied. A single chip consists of millions of gates.

GDSII—Graphical Design System II. An industry standard format for exchanging final IC physical design data between EDA systems and foundries or mask makers. GDSII is a Cadence standard.

GSM—Global System for Mobile communications. World's first standard for mobile communications.

GUI—Graphical User Interface.

Hard IP—Complete description of the circuit at physical level. Hard IP is routed, verified, and optimized to work within specific design flows.

HDL—Hardware Description Language. A high-level design language in which the functional behavior of a circuit can be described. VHDL and Verilog are HDLs that are widely used.

HDL-A—Hardware description language for describing analog designs.

HW/SW—Hardware/Software.

HW/SW co-design—Design methodology that supports concurrent development of hardware and software to achieve system functionality and performance goals.

HW/SW co-simulation—Process by which the software is verified against a simulated representation of the hardware prior to system integration.

HW/SW co-verification—Verification activities for mixed hardware/software systems that occur after partitioning the design into hardware and software components. It involves an explicit representation of both hardware and software components.

IACK—Interrupt Acknowledge.

IC—Integrated Circuit.

ICE—In-Circuit Emulator.

IEEE—Institute of Electrical And Electronic Engineers.

IEEE-1284—Standard for personal computer parallel ports.

IEEE-1394—High-speed serial bus. Also called a firewire.

I/O—Input/Output.

IP—Intellectual Property. IP is the rights in ideas that allow the owner of those rights to control the exploitation of those ideas and the expressions of the ideas by others. IP includes products, technology, software, and so on.

IR Drop—Current-resistance drop.

IRQ—Interrupt Request.

ISR—Interrupt Service Routine.

ISS—Instruction Set Simulator.

JPEG—Joint Photographic Experts Group. Industry standard for the digital compression and decompression of still images for use in computer systems.

JTAG—Joint Test Access Group. IEEE 1149.1 standard for device scan.

LA—Logic Analyzer.

Layout—The process of planning and implementing the location of IC devices within a chip design.

LEF—Library exchange format

Logic BIST—Logic Built-In-Self-Test.

LVS—Layout Versus Schematic.

Manufacturing test—Physical process of validating and debugging the performance and functional operation of semiconductor chips/products.

Micron—One-millionth of a meter, or about forty-millionths of an inch (0.000040 inches)

MPEG—Moving Picture Experts Group. Industry standard for the digital compression and decompression of motion video/audio for use in computer systems.

MUX—Multiplexor.

NC-Verilog—Simulation tool available from Cadence.

Netlist—Complete list of all logical elements in an IC design, together with their interconnections.

N2C—System design solution available from CoWare.

OPC—Optical Proximity Correction.

OVI—Open Verilog International, a Verilog HDL standard body.

PCB—Printed Circuit Board.

PCI—Peripheral Component Interconnect bus.

PE—Parasitic Extraction.

PLD—Programmable Logic Device.

PLI—Programmable Language Interface.

PLL—Phased Locked Loop.

Process—Steps by which the ICs are constructed for a given technology.

PROM—Programmable ROM. ROM that can be programmable by the user.

PRPG—Pseudo-Random Pattern Generation.

Protocol—Formal definition of the I/O conventions for communications between computer systems and peripherals.

Prototype—Preliminary working example or model of a component or system. It is often abstract or lacking in some details from the final version.

PSM—Phase Shift Mask.

RAM—Random Access Memory.

RE—ROM Emulator.

RF—Radio Frequency.

RISC—Reduced Instruction Set Computer.

ROM—Read Only Memory.

RPS—Rapid Prototyping System.

RTL—Register-Transfer Level.

RTOS—Real Time Operating System, such as VxWorks, pSOS, or Windows CE.

SDF—Standard Delay Format.

Semiconductor manufacturer—A firm that is active in the business of designing and producing semiconductor devices.

SI—Signal Integrity.

Simulation—Simulating a chip design through software programs that use models to replicate how a device will perform in terms of timing and results.

SOC—System-On-a-Chip. An IC that contains the functional elements of an entire electronic system, such as a computer, PDA, or cell phone. SOC designs involve integrating CPU, memory, I/O, DSP, graphics accelerators, and other components on a single chip.

Soft core—Soft core is delivered in the form of synthesizable HDL code.

Specman Elite—A testbench-generation tool available from Verisity.

SpectreHDL—Cadence hardware description language that describes analog designs.

SPF—Standard Parasitic Format.

SPICE—Simulation Program With Integrated Circuit Emphasis.

SPW—Signal Processing Worksystem. Signal processing system design tool available from Cadence.

SRAM—Static Random Access Memory.

STV—Static Timing Verification.

TBV—Transaction-Based Verification.

TCL—Tool Command Language.

TestBuilder—Testbench authoring tool available from Cadence.

TRST—Transaction Recording System Task calls.

TTM—Time to Market.

TVM—Transaction Verification Model.

UART—Universal Synchronous Receiver Transmitter.

USB—Universal Serial Bus.

VCC—Virtual Component Co-design. System design tool available from Cadence.

Vera—Testbench generation tool available from Synopsys.

Verification—Pre-silicon process that is used during the design phase for gaining confidence that the design will produce the expected results.

Verilog—Industry-accepted standard HDL used by electronic designers to describe and design chips and systems prior to fabrication.

Verilog-A/MS—HDL for describing analog/mixed signal designs.

Verilog LRM—Verilog Language Reference Manual.

Verilog-XL—Verilog simulator available from Cadence.

VHDL—VHSIC Hardware Description Language.

VHDL-AMS—Hardware description language for describing analog/mixed signal designs.

VHSIC—Very High Speed Integrated Circuit.

VC—Virtual Component. A design block that meets the VSI Specification and is used as a component in the virtual socket design environment. Virtual components can be available in three forms: soft, firm, or hard.

VCS simulator—A simulation tool available from Synopsys.

Virtual Prototype—Computer simulation model of a final product, component, or system.

VSI—Virtual Socket Interface. Set of standards to enable the exchange of IP building blocks. VSI is supported by the VSI Alliance, a 148-member group that was formed to address the complex problem of establishing comprehensive standards for the exchange of IP components between semiconductor companies.

Index